MIGRATION
paths through time and space

Migration: paths through time and space is part of the Biological Science Texts Series catering for a wide range of biological interests. Some of the titles in this Series are given below. A complete list is available from the publisher.

The Evolutionary Ecology of Animal Migration
R. Robin Baker

The Genesis of Diversity
Bryan Shorrocks

The Study of Man
E. J. Clegg

Human Navigation
and the sixth sense
R. Robin Baker

Sociobiology and Behavior
David P. Barash

MIGRATION
paths through time and space

R. ROBIN BAKER
Department of Zoology, University of Manchester

HM HOLMES & MEIER PUBLISHERS, INC.
IMPORT DIVISION
IUB Building
30 Irving Place, New York, N.Y. 10003

> *To the memory of*
>
> *HOWARD HINTON.*
>
> *"It's a good sign if nobody believes you — as long as you are right."*

British Library Cataloguing in Publication Data

Baker, R. Robin
   Migration.
   1. Animal migration
   2. Evolution
   I. Title
   591.52   QL754 . B3

ISBN 0 340 26079 3

34,465

Typeset by Macmillan India Ltd., Bangalore.
Printed in Great Britain for
Hodder and Stoughton Educational Limited,
Mill Road, Dunton Green, Sevenoaks, Kent,
by Richard Clay (The Chaucer Press) Ltd Bungay, Suffolk.

Preface and acknowledgements

Many people consider 'behavioural ecology' to be simply a new name for ethology. I disagree. In my view there is a real and revolutionary difference between the two. Nowhere is this difference more marked, and more important, than in the study of migration.

Four years ago I published a large book entitled '*The Evolutionary Ecology of Animal Migration*'. That book was an attempt to do many things. Not least, it was an attempt to produce a textbook that reviewed developments since the last comparable attempt by Heape (1931) and to counteract the parochialism that had increasingly characterised the study of animal migration during the intervening half century. More than that, however, the book was an attempt to apply the behavioural ecologist's approach to a subject which had been well and truly the domain of ethologists for over 20 years.

I have received many reactions to the book from academics. Some reactions have been most favourable, and for these I am grateful; others have not. Although it is a little early to be sure, a pattern seems to be emerging in that, on the whole, behavioural ecologists have received the book much more favourably than ethologists. As far as my own undergraduates are concerned I detected an initial, fairly subtle, reluctance to accept immediately the interpretation of migration that I advanced in the large book. More than for any other reason, this is because, if at school they studied behaviour at all, they were raised in biology departments steeped in the ethological tradition. With such a background, the idea that animals explored, made judgements, and in general solved their spatial and temporal problems in much the same way as Man was alien to the concept of animal behaviour around which they were taught. Yet this anthropomorphic concept of animals was implicit in the familiar area model of animal movements which was one of the major theses of my previous book.

The anthropomorphic concept, however, was there only by implication. It arose inevitably, but was never stated, as the familiar area model itself evolved from a detailed consideration of animal movements, particularly of vertebrates. In the present book I have made the consideration of this concept one of the major themes. At the same time, I advance the view that behavioural ecology is a new ideology, completely different from ethology. It is still interested in the innate predispositions on which all behaviour is based. Indeed, the unravelling of the evolution of these predispositions is one of the major stated aims of the subject. Unlike ethology, however, behavioural ecology is not reductionist. No longer are animals seen only as bundles of innate reflexes, travelling from one automatic response to another. Instead, at the level of the individual, a qualitatively different picture emerges: that of a sentient organism, travelling through time and space, attempting as it does so to solve all the problems thrown at it by its environment.

This vision of the animal as an individual is particularly important to an understanding of animal migration, the study of which has for so long been torn apart by on the one hand the ethologist's search for programmed reflex responses and on the other the population ecologist's search for mathematical means of describing dispersion, gene flow, and the regulation of population density. The only level that was largely ignored was that of the individual actually performing the movements.

This book, then, is not a mini–version of my previous text-book, although there may be a superficial resemblance in that much of the basic information on animal movement patterns is conveyed using the diagrams and illustrations from the earlier volume. There the similarity ends. Instead, the information compiled in the previous book and the models that were first formulated therein are here used as a springboard for presenting to undergraduates and their teachers a basis for the study of animal behaviour in general. Using animal movement patterns as the theme, an attempt is made to spell out the importance and true nature of the behavioural ecology revolution. The importance of studying animal behaviour at the level of the individual is stressed and it is shown how much of the conceptual legacy from the age of ethology may have to be abandoned before the revolution can be considered to be complete. The text argues also that anthropomorphism is no longer the sin that previous generations of behaviourists have considered it to be. The main aim of the book is to provoke discussion and argument, not just about migration, but also about the concept of the organism on which the whole study of behaviour is based.

The approach used in this book is due to a suggestion by Dr. Edward Broadhead of Leeds University and I am once again indebted to him for making me realise the potential of such an approach. I have also unashamedly plaguerised many words and phrases from among the notes and comments that he made on an early draft of the manuscript. His perception and diligence as Biological Editor have improved beyond measure the final text as compared to its original form.

The philosophy presented in this book was sharpened considerably by many long and detailed discussions with Janice G. Mather. I am grateful to her, both for those discussions and for allowing me to use some of her unpublished results. I also thank those undergraduates that have allowed me to use their unpublished work.

It has once again been a pleasure to work with the editorial, design and production teams at Hodder and Stoughton. It was worth the trauma of producing another manuscript to resume such an enjoyable and productive relationship.

My final acknowledgement, however, must again be to my wife and family. Once more they were prepared to pretend I did not exist and to live their lives around me while I got this fourth book out of my system. Were they less tolerant than they are, I should still be only part-way through the first book.

Manchester   Robin Baker

Contents

1

Migration and the behavioural ecology revolution

The study of animal behaviour has just emerged from a period of revolution. *Behavioural ecology*, the evolutionary study of all forms of behaviour, is the new regime. It emerged during the 1970s as the product of a marriage between population biology, behaviour and ecology, and was carried to prominence on the back of its most flamboyant sub-discipline: *sociobiology*, the evolutionary study of social behaviour.

Behavioural ecology has taken the place of *animal psychology* and *ethology* as the new establishment for behavioural theorists. All aspects of animal behaviour are in the process of being brought into line with the new ideology and the behavioural study of migration is no exception.

In the early years of study into behaviour, animal psychologists such as Pavlov and Skinner saw the behaviour patterns of vertebrates as chain reflexes shaped by a history of reinforcement episodes. Man's behaviour, in particular, was seen as the totally plastic product of environmental influences. Ethology, on the other hand, grew up out of the work of the early naturalists, such as Gilbert White, Jean Fabre, and Charles Darwin, and came of age when people such as Lorenz, Tinbergen and von Frisch took the scientific study of behaviour out of the laboratory and into the field. Ethologists emphasised the innate aspects of behaviour, argued that behaviour patterns could be used to identify phylogenetic relationships between animals, and sought adaptive values for the behaviour patterns that they described (Parker 1978).

The study of migration took on the mantle of ethology with enthusiasm. Before the behavioural ecology revolution an animal was seen as an automaton, a very simple machine. Put in a stimulus at one end and an invariable output appears at the other. A little learning may be involved, but this is either immediate and irreversible (i.e. *imprinting*) or the product of repeated trials and mistakes (i.e. *conditioning*). Some allowance was made for several stimuli to be involved, but generally the approach was reductionist. Expose a bird to a particular photoperiod and show it the Sun and it will automatically fly south. Place an adult salmon in a current of water containing a chemical on which the fish was imprinted when young and it will swim irresistably upstream. The view is simple, straightforward, uncluttered and *inflexible*.

Contrast this with the behavioural ecologist's concept of an animal as an individual confronted by dangers and problems that have to be solved if the animal is to survive and reproduce. Always the animal has several courses of action (i.e. *strategies*) available to it. Which strategy it adopts in solving any particular problem set by the environment will depend on its genetic predisposition and past experience.

Each individual is a package of predispositions, often a unique package, a sample of all those available within the gene pool. These predispositions form the major weapon with which an individual embarks on life. Over the generations they have been tried and tested against the uncompromising yardstick of natural selection. Furthermore, they have survived; not only have they survived but the chances are that the individual has inherited those predispositions that have survived in the greatest numbers. Most individuals should be well-disposed, therefore, to cope with all the problems that the environment should set during their lifetime.

For many animals, one of the most important predispositions is the ability to learn; but here we encounter one of the major internal conflicts of the new regime. Many behaviourists believe that although animals learn from experience they do so according to certain rules, rules based on genetic predispositions to interpret the environment in a particular way. There is some evidence for this. For example, birds have to learn the pattern of movement of the Sun across the sky during the day before they can use it to determine compass direction. They seem predisposed, however, to learn that the Sun does move across the sky. It is more difficult to teach a bird that the Sun does not move (Matthews 1955).

Where does Man fit into this picture? Many social scientists would like to believe that humans are born without genetic predispositions to behave in particular ways. In consequence, they see all human behaviour as environmentally determined. Many sociobiologists, however, consider that all human behaviour is like that of a bird and the Sun, influenced by genetic predisposition even if it is possible to learn to behave in a completely different way. Whichever view is correct, there are no grounds for believing that Man is qualitatively unique in the way that an individual's past experience interacts with genetic predisposition to produce observed behaviour. In the absence of such grounds, the same approach and yardstick has to be applied to all animals. If Man is a sentient creature, we have to allow other animals also to be sentient. If other animals are not sentient, then neither is Man.

So here we have the behavioural ecologist's view of an animal: as an individual; a package of predispositions and memorised experiences; the product of natural selection; a sentient creature prepared to solve any problem that a complex environment can set or to die in the attempt.

One of the problems that faces all animals is where to live. Individuals have little control over where or at what stage of development they first appear. Parents decide these things according to their own best interests. Thereafter, the individual has to wend its way through time and space until it dies. The most successful individuals in terms of the number and quality of offspring they produce (i.e. *reproductive success*), will be those with a path that exploits the environment to the full, minimising costs, maximising benefits, always seeking the best trade-off between the two. The path that an animal produces is its *lifetime track*, the outward manifestation of the individual's solution to spatial and temporal problems, the playing off of inherited predispositions and acquired experience against the environmental backcloth, running the gauntlet of natural selection for yet another generation.

The study of animal migration is the study of the lifetime track. Chapter 2 looks at previous models of these tracks, our legacy from the age of ethology, and asks how much of this legacy should be retained and how much rejected. The conclusion is that behavioural ecology should begin its adoption of the study of migration with a major purge of the old ideas. A new start and a new direction are required. The first step in this has to be the attainment of perspective over the whole range of animal movements, seeking similarities and differences in the lifetime tracks of animals from widely different taxonomic groups. When this step is taken (Chapter 3), an important theme emerges—a similarity between the lifetime track of Man and a wide range of other animals. This similarity is probed in Chapter 4 and the question is considered whether Man is after all unique in possessing a cerebral sense of location; perhaps other animals know where they are and where they are going; perhaps other animals solve their spatial and temporal problems by thinking. Indeed, once this possibility has been contemplated, it becomes difficult to find animals that do not have a cerebral sense of location. One possible solution is that the mechanisms used by other animals in gaining their sense of location are different in some way from those used by Man. Chapters 5 to 8 consider from this viewpoint the most important of these mechanisms: exploration. They conclude that exploration is an all-pervading behavioural element in animals that have a sense of location and that move short distances. The final challenge to this pervasiveness comes from long-distance seasonal migrants. This challenge is considered in Chapter 9, which concludes there is no fundamental difference in the mechanisms used by the classical migrants, such as birds and fish, and the more 'sedentary' animals, such as modern Man. Having reached this conclusion, we then consider in Chapter 10 how this affects our understanding of that most elusive of all migration mechanisms: *navigation*, the means by which animals find their way to a required, usually familiar, destination across unfamiliar terrain. The picture that emerges is that if navigation is considered as an element of exploration, rather than purely of long-distance migration, new light can be shed on the nature of the navigational mechanism itself.

Chapter 11 suggests that it is time to create a new paradigm: that we assume all animals explore and have a cerebral sense of location unless there is evidence to the contrary. The remainder of Chapter 11 and Chapters 12 and 13 pursue such evidence, seeking animals with different ways of life. Some are found but many question-marks remain.

The book ends by returning to the major issue of contention within behavioural ecology. Chapter 14 asks how much of the migration behaviour of Man and other animals is innate and how much is learned. It asks, also, what relevance the study of the lifetime tracks of other animals may have to an understanding of human behaviour and what insight into the migrations of other animals can be gained from a study of Man.

No new regime can totally ignore the legacy from the past. There are always lessons to be learned; mistakes to be avoided or rectified; and sometimes a few ideas useful enough to be retained, albeit stamped by the new ideology. In any case, the behavioural ecology revolution is not yet complete. Pockets of resistance remain and there still exist, not least among students of migration, behaviourists who cling to the old ideas. The path to modern migration theory may not lie in the past, but an understanding of past perspectives and ideas may help us to assess the correctness of the new ideology.

2

Legacy from the past

As the study of animal migration prepares to move into the age of behavioural ecology, it brings with it from the past a legacy in three parts: words, concepts and facts.

Behavioural ecology is very particular about (reactionaries would say preoccupied by) the words that people use. Richard Dawkins and John Krebs in a lecture to the Royal Society stressed that the words scientists use to describe their subject influences in no small measure the way that they think about it. This influence is insidious, all the more so because scientists have a history of dismissing words as 'mere semantics'. However, there should be no doubt about the power of terminology. A clear and appropriate terminology can be a subject's greatest strength; a confused and inappropriate terminology its greatest weakness. As it stands, poised at the start of a new era, the study of migration falls into the latter category.

For two thousand years, discussions of animal migration centred on birds and on an explanation for their seasonal appearance and disappearance (Baker 1980 b). All this time debate raged: did some birds migrate south for the winter or did they instead transmutate into other species, hibernate in mud, or fly to the Moon? Aristotle had been convinced that birds did migrate but it was not until the end of the nineteenth century that this view became generally accepted. Not all birds were migratory, of course; some failed to disappear for the winter and thus were non-migratory. A natural division was evident. Birds were either migratory or resident; migrant or non-migrant.

Two thousand years of debate had produced an indelible axiom in peoples' minds: birds migrate; migration is what birds do. In an influential book, Heape (1931) made a bold attempt to find examples of other animals that were 'true' migrants (i.e. behaved like birds). Some, such as fish and whales, were accepted as migratory. The remainder, like resident birds, were non-migrants. Nevertheless, Heape observed that some of the non-migrants also showed interesting movements that deserved attention and suggested two further categories of track pattern to rank alongside that of migration. These were *nomadism* and *emigration*. Movements that were of lesser interest had previously been termed *trivial* (Pearson and Blakeman 1906).

Emigration implies movement away from unsuitable conditions. It soon came to be realised that such movements usually produced colonisers as well as refugees. For these, the term emigration was inadequate. In retrospect, the term eventually adopted in its place was equally inadequate and even more misleading. Emigration had usually been applied to situations in which animals scattered in all directions from their place of origin. Because of this, a term that had often been used in connection with emigration was *dispersal*, and eventually the former was displaced by the latter.

The first major terminological niche for dispersal was perhaps among mammalogists. It was used primarily to describe the behaviour of young mammals when they abandon the area in which they were born and settle down elsewhere. Ornithologists also adopted the term to describe a similar phenomenon: the movements of birds just after fledging. The term was also applied to adult birds and mammals to describe a change in breeding site which often also involves a scattering of previous breeding groups. By the time Dorst (1962) published his important book on bird migration, the terminology was established. All non-trivial movement was either migration, dispersal or nomadism. Put another way, everything that was neither migration nor nomadism was dispersal. Rapidly the literal meaning of the term was forgotten and it became involved in all manner of semantic corruptions.

The literal meaning of the verb 'to disperse' is to scatter, and when first adopted this was more or less the sense in which it was used. In the mind's eye at least, though not necessarily in reality, juvenile birds and mammals scatter from their place of birth. However, it was often also necessary to refer to the behaviour of individuals. At first, groups were 'showing dispersal' and the individuals were 'dispersing'. So, what does an individual do? Obviously, it disperses! The corruption had begun (an individual cannot, of course, disperse, except by exploding). Worse was to follow in the hands of entomologists.

The major early work on insect migration was by C.B. Williams (1958). Williams saw insect migration as directed and as involving to-and-fro seasonal reversals, often on a north–south axis. He concluded that insects migrate. Eleven years later, C. G. Johnson (1969) argued that insect movements were not directed. It followed that the behaviour Williams had described as migration was really dispersal. With total disregard for its literal meaning, Johnson proceeded to apply the term to all non-trivial insect movements, whether by groups or by individuals. The final corruption came when the *convergence* upon hibernation and aestivation sites of vast numbers of some species of insects was included under the general umbrella of *dispersal*.

Johnson engineered one further twist of terminology that was symptomatic of the parochialism that was developing among students of migration. Having decided that insect migration was really dispersal, he did not then reverse his logic and decide, as an ornithologist would have done, that if the movement was dispersal it could not also be migration. Instead, Johnson saw the terms as synonymous, and went on to define migration as the 'adaptive change of breeding habitat'. Such a definition was in the same spirit as the use of the term dispersal by mammalogists, and ornithologists. By 1970, ornithologists and most students of vertebrates used the term migration to refer to a to-and-fro return movement that ended where it began: at the original breeding site. Entomologists used the term in precisely the opposite sense: to indicate a change in breeding site.

While ornithologists and entomologists were developing their own, diametrically opposite, uses of the term migration, a third group was also going its own way. Quite by chance, demographers among anthropologists and social scientists were developing a terminology to describe human movements that allied them to entomologists rather than to students of other vertebrate groups. Migration and dispersal were used more or less synonymously, but the emphasis was different. Whereas entomologists were inclined to say that migration is really dispersal, students of human demography were inclined to say that dispersal is really migration. As the final irony, humans that perform seasonal return migrations comparable to those of birds were described, not as migrants, but as nomads.

When the behavioural ecology revolution began, therefore, the study of migration was encumbered with a terminology that was semantically corrupt, parochial, and downright contradictory. If Dawkins and Krebs are correct that the words scientists use to describe their subject influence in no small measure the way that they think about it, there would seem little hope of further progress in migration without a major semantic purge.

Most of the early arguments concerning migration were terminological. Concepts were slower to develop, migration behaviour being explained away initially under the simple umbrella concept of 'instinct'. Toward the end of the zenith of the animal psychologists, Fraenkel and Gunn (1940) published an influential book in which movements were classified on a physiological basis according to the nature of response to environmental stimuli. Much of Fraenkel and Gunn's terminology (taxes and kineses, etc.) has survived. Their terms have the insidious effect, however, of stamping animals as automatons. Present a stimulus at one end and an invariable orientation response appears at the other.

In the 1950s, the emerging discipline of ethology, with its emphasis on innate behaviour, received, as a legacy, 2000 years of discussion over whether animals, particularly birds, were migrants or non-migrants and the (by then classic) view that orientation responses were simple reflexes. The conclusion reached was inevitable. Migrants were animals with particular and precise innate orientation responses that were lacking in non-migrants. One of these innate responses was navigation, and the most influential book of the decade was that of G.V.T. Matthews (1955) on bird navigation. In essence it examined ways in which birds could find their way from one place to another using a set of simple innate responses to the Sun's arc.

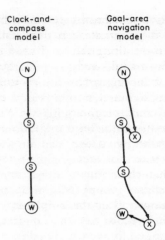

Fig. 2.1 The clock-and-compass and goal-area navigation models of the autumn migration of young birds
N = natal site; S = stopover site; W = 'winter' site.
In the clock-and-compass model the bird flies in an innately programmed direction for an innately programmed length of time before stopping. Repetition for an innately programmed number of times results in the bird arriving in suitable winter quarters. In the goal-area navigation model, the coordinates of each S and W are encoded within the bird's genetic material. If, for some reason, the bird fails to arrive at S but lands instead at X, at the first opportunity it migrates back to the appropriate position.
[Modified from Baker (1978a)]

The age of ethology produced two major models of bird migration (Fig. 2.1). The first was the 'clock-and-compass' or 'one-direction orientation' model derived primarily from experiments carried out by Perdeck (1958). According to this model, a juvenile bird migrating for the first time inherits a programme whereby it migrates in a particular direction for a programmed length of time. After feeding-up and refuelling, the process is repeated. After a programmed number of repetitions, the bird eventually ends the programme in appropriate winter quarters. Before having set out on its first autumn migration the young bird is supposed to imprint on some characteristic (perhaps related to the Sun's arc) of its breeding site. While in its winter quarters, a similar imprinting process is supposed to take place. The return migration in spring is partly a reversal of the clock-and-compass autumn programme and partly navigation back to the imprinted breeding site. Thereafter, as an adult, all migration is performed by navigation across unfamiliar territory to the now known breeding and winter quarters.

Many ornithologists disliked the clock-and-compass model because it seemed so sensitive to the vagaries of wind and weather. Rabøl (1970) suggested an alternative, the goal-area navigation model, that overcame this problem and that incidentally was a harder-line ethological view. According to this model, birds have programmed within their genetic material the geographical coordinates of where they should be at any particular time of year. For migrant birds, these coordinates are programmed to slide back and forth along the traditional migration route as the annual cycle unfolds. Both adults and juveniles, therefore, navigate their entire migration route to programmed destinations.

Whichever model was favoured, the concept of migration behaviour in the minds of ethological ornithologists was clear. Birds were either genetic migrants or genetic non-migrants. Migrants had need of navigation, non-migrants did not. Migrants were therefore programmed to be able to navigate, non-migrants were not. It says a great deal for the power of the ethological climate that such a view of migration and navigation could arise and fixate in the face of an enormous paradox. Most of the experiments on bird navigation involved Homing Pigeons, the domesticated descendants of the Rock Dove, *Columba livia*. As far as the establishment was concerned, the Rock Dove is non-migratory. Hence it has no need of navigation and must lack the necessary innate reflexes. People predisposed to accept the results of experiments on Homing Pigeons argued that stringent selection by Man for good homers among pigeon stocks had produced birds genetically programmed for such reflexes. Others maintained that Homing Pigeons could not have the same genetic programmes for navigation as those possessed by true migrants and that results obtained for Pigeons gave no insight into the navigation of migratory birds.

Students of other vertebrate groups also sought ethological models to explain migration behaviour, but of these only students of *anadromous* fish and sea turtles were successful. Anadromous fish are those that spawn in freshwater but grow to maturity and feed as adults in the sea. Examples are salmon, lampreys and sturgeon. Sea turtles lay their eggs in holes dug on sandy tropical beaches but feed elsewhere, often thousands of kilometres away. It was suggested that both groups of animals imprint when young on the chemical 'signature' of the water around their hatching site. They then migrate downcurrent; fish to the sea, turtles to feeding grounds. Navigation back to the breeding grounds involves encountering water carrying the chemical on which they have imprinted, whereupon up-current orientation and swimming is automatically triggered (Hasler and Wisby 1951, Carr 1972). For those Green Turtles, *Chelonia mydas,* that breed on Ascension Island and feed along the coast of Brazil, this story was embellished further in relation to the process of Continental Drift (Carr and Coleman 1974). Ascension Island Green Turtles are considered to be an inbred lineage with an ancestry going back to a time before the formation of the southern Atlantic Ocean. As the ocean first formed, ancestral Green Turtles are considered to have travelled the short distance between feeding grounds on the coast of the new Brazil and the first of the volcanic piles that are a frequent feature of mid-oceanic ridges (Fig. 2.2). Carr and Coleman envisage that the ancestral turtles may have evolved an innate response to

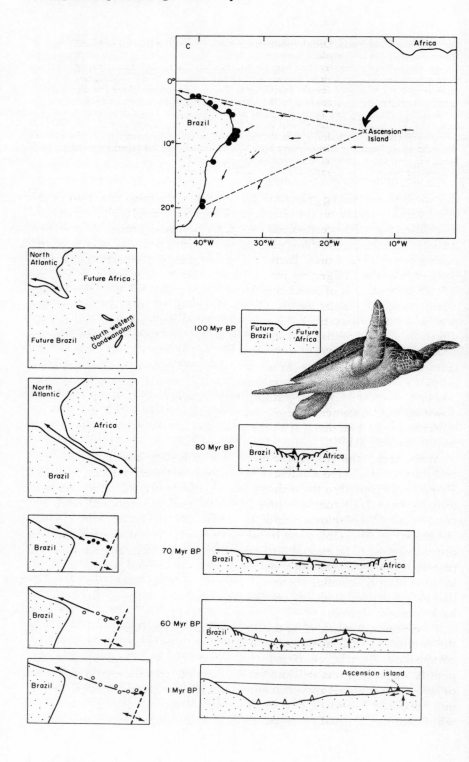

Fig. 2.2 The Continental Drift hypothesis for the origin of the Ascension Island deme of Green Turtles, *Chelonia mydas*
Female Green Turtles that lay their eggs on Ascension Island in mid-Atlantic spend two or three years between each visit feeding along the coast of Brazil. In the top diagram the dots show recaptures of tagged female Turtles from Ascension. Arrows show the direction of surface water currents. According to the Continental Drift hypothesis, ancestors of this deme both nested and fed along the coast of north-western Gondwanaland. With the formation of the South Atlantic, this deme extended its migration axis (double-headed arrow) to breed on the volcanic pile that appeared in the mid-ocean rift. As the Ocean widened, new 'Ascension Islands' were born, old ones became submerged (open circles and triangles), and the migration track gradually lengthened.
[Compiled from Baker (1978a). after Carr and Coleman]

---

migrate from the feeding grounds in the direction of the rising Sun. This would cause them to encounter the olfactory plume that would spread downcurrent from the mid-ocean island. As the continents drifted further and further apart, extinct volcanic piles became submerged only to be replaced in mid-ocean by new ones. Each time this happened, the turtles could find the new island by migrating just a little further in their innate direction.

The early models of insect migration propounded by Williams (1930, 1958) required similar levels of involvement of the animal in the production of its own track (i.e. adoption and maintenance of a particular direction). In the 1950s and 1960s, however, there began a movement to take away even this level of involvement of an insect in its own track. The culmination of this movement was a major publication by C. G. Johnson (1969). He argued that whereas an insect undoubtedly controls its own take-off, thereafter its migration track is primarily the product of downwind displacement. Large, strong-flying insects such as locusts and butterflies were considered as unlikely to influence their own track as small insects, such as aphids.

At about the same time as the role of insects in their own migration was being denigrated, so too was that of oceanic fish. In the early 1960s, Saila and Shappy (1963) tried to show that salmon could return from their oceanic feeding grounds to coastal waters (there to pick up imprinted olfactory cues) simply by employing random search. Later, Harden Jones (1968) elaborated a comparable view by suggesting that the migration circuits of oceanic fish could be executed by passive drift. All surface ocean currents resolve into closed rings of gyrals and for any current at one depth or location there is a counter-current at some other depth or location. Any fish that passively drifts with an oceanic water current would eventually arrive back where it started.

To different degrees, all of these models considered that, at the level of the individual, migration behaviour is stereotyped and inflexible. To compensate for this, an element of flexibility was introduced at the population level. It was well known, for example, that the migration tracks of juvenile vertebrates are much more variable than those of adults. Young individuals are liable to appear in a much wider variety of places, many of which are not normally visited by adults (Fig. 2.3). According to the

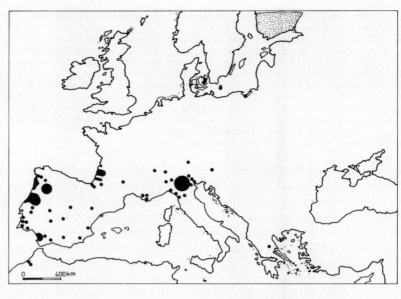

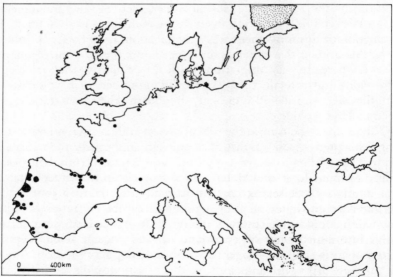

Fig. 2.3 Recoveries of Pied Flycatchers, *Muscicapa hypoleuca*, ringed in Finland in the breeding seasons 1948–67

The upper figure shows the returns of birds ringed as pulli in the nest and recovered during their first autumn. The lower figure shows the returns of older birds that have performed a complete annual migration cycle on at least one occasion. The young birds are much wider ranging than the adults. Such data have been interpreted to show strong selection against the birds on the flanks (e.g. Italy) of the traditional migration route (but see Chapter 9) [Redrawn with modifications from Rabøl (1978) after ringing reports in *Mem. Soc. Fauna Flora Fennica*. Photo by Eric Hosking]

established models, it would be usual for juveniles with tracks not the same as adults to be selected against in each generation (Rabøl 1978).

Occasionally, however, such deviant juveniles may provide the genetic raw material from which a new migration route could become established. Similarly, among insects there are the 'lucky few' that the wind deposits by chance in a suitable rather than an unsuitable habitat. As insects are such prolific breeders, the unlucky remainder are viewed as expendable (Johnson 1969).

These are the major elements in the conceptual legacy from the past. At best, migrant animals are automatons. Genetically different from non-migrants, they are bundles of innate reflexes, programmed to respond in a stereotyped, inflexible way to sequences of environmental stimuli. Modern migration routes are the result of natural selection acting over the generations to favour those programmes of reflexes that result in such routes. When some element of learning is unavoidable, such as in the recognition of a home site, the learning process is the simplest possible(i.e. imprinting) and occurs at a genetically programmed receptive stage. At worst, the migrant is seen as a virtually inanimate particle, like a wind- or water-borne seed, travelling through space in a sensory void. At the individual level, migration behaviour is seen as inflexible. Indeed the individuals themselves are expendable.

Such a legacy is clearly alien to the behavioural ecology concept of an animal as a sentient creature, attempting as an individual to solve the spatial and temporal problems created for it by its environment. Yet we have to be careful with this legacy. Much of it is based on irrefutable fact and cannot be dismissed in its entirety as can the terminology that comes with it. Some migrant birds really are predisposed to fly in a particular direction. Salmon really do smell their way home. Aphids really are carried along by the wind. These facts must be accepted, along with many others. The question we, as behavioural ecologists, have to ask ourselves is whether the facts are even more compatible with the new ideology than they were with the old. If they are, we can accept the facts as a legacy but reject the conceptual framework of which they were a part.

The study of animal migration brought to behavioural ecology for acceptance a massive legacy of terminology, fact, and conceptualisation. The facts have to be accepted, but little else is useful. Animal migration enters the new age needing a new terminology and a new conceptual framework. Most of the problems encountered by the old framework can be traced to its early preoccupation over which animals are migrants and which are not. We should begin, therefore, by looking at the movements of all animals, not only those that are classically considered to be migrants.

3
Lifetime tracks: paths from birth to death

The lifetime track is an individual's solution to the spatial and temporal problems that it encounters in its environment. It is a reflection not only of the environment but also of the individual's predispositions, given substance and often modified by past and present experience. If *migration* is defined as the act of moving from one spatial unit to another (Baker 1978a), then a lifetime track is made up of all the migrations performed by an individual between birth and death. Lifetime tracks are the raw material for the study of animal migration.

The first step in any behavioural study is to marshal the facts, to identify the range of behaviour that awaits explanation, and most important of all, to determine perspective to see the most important and interesting questions but only in the context of all questions. In the case of migration, a 'feel' has to be acquired for the variety of ways that animals thread their way through time and space on their way from birth to death.

Consider your own lifetime track. What is its most regular feature? The last migration you make each day takes you to a place to sleep. More often than not, this migration will be a *return* migration (i.e. to a place visited previously). Moreover, it will most often be a return to the place you slept in the previous night. What is your age in days? What proportion of those days did you return at night to the place you slept in the previous night? You will be an unusual human indeed if the second figure is not well in excess of 90 per cent. Like a great majority of your conspecifics, the major feature of your lifetime track is a *daily return migration cycle* that ends at night where it began the previous morning. This daily cycle is the building block of your lifetime track; so it is for most humans.

Man may be divided into four major ecological groups: (1) hunter–gatherers, (2) pastoral nomads, (3) agriculturalists, and (4) industrialists. Hunter–gatherers have no domesticated animals (except dogs), cultivate no crops, and subsist on 'large' animal food (obtained by males by hunting or fishing) and fruit, roots, grain and 'small' animals (obtained by females and young by gathering) (Lee and DeVore 1968). Pastoral nomads cultivate no crops but herd one or several species of ungulates, accompanying them throughout the latter's annual cycle (Krader 1959). Agriculturalists cultivate crops and usually practise animal husbandry (Barnett 1971). Industrialists also subsist on cultivated plants and domesticated animals but are characterised by only a small proportion of each *deme* (i.e. group of individuals; as in *demography*) actually producing food, the majority being engaged in other activities, the products of which are then traded for food. Virtually all industrialists and agriculturalists have a relatively permanent sleeping site in a house or hut. For much of the year this is true also for pastoral nomads, many of whom sleep in tents, and for a variety of hunter–gatherers.

Although there are exceptions, the daily return migration cycle is the most regular feature of the lifetime track for the majority of these humans and the same is true for many other animal species. Even if they are not asleep, most animals have one or more resting periods during the course of each 24 hours. Usually a particular type of site is chosen in which to rest (or roost) and often, as in the case of Man, the same location is used several days or nights in succession. Non-human primates may return each night to sleep in a favoured group of trees (e.g. Baboons, Fig. 3.1). Rodents may build nests or excavate burrows to which they return to sleep or shelter.

Fig. 3.1 Track patterns of the Patas Monkey, *Erythrocebus patas*, (solid line) and the Savanna Baboon, *Papio cynocephalus*, (dashed lines) at Chobi, Murchison Falls National Park, Uganda. Solid dots show overnight roosting sites
[From Baker (1978a), after Hall]

Bats return, after foraging, to the same cave, building or hollow tree. Bees return each evening to their hive. As the tide falls, inter-tidal fish may shelter day after day in the same rock pool while limpets return to the same crevice or depression in the rock; and so on. For all of these animals, the daily (or tidal) return migration cycle is the major building block of their lifetime track.

In order to sleep, humans return each night not only to the same building as the night before but usually also to the same bed. *Degree of return* is high. For most of the animals just mentioned, return to the roosting site is equally as precise and frequent as for Man. Others, however, seem less fussy. Thus,

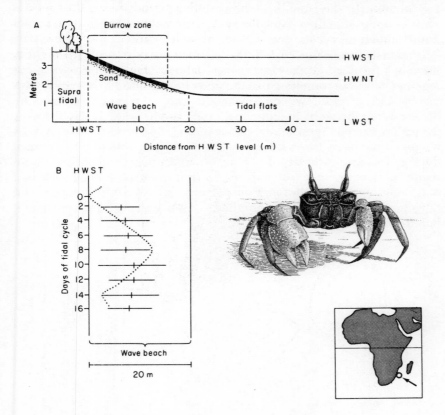

Fig. 3.2 Daily/tidal and semi-lunar return migration cycles of the Ghost Crab, *Ocypode ceratophthalmus*, on Inhaca Island, Moçambique

Ghost Crabs on Inhaca Island spend the day buried in the sand of the wave beach (A) and emerge and become active at night during periods when they are not covered by the tide. They feed, either on the wave beach or on the tidal flats. As the tide comes in once more or as dawn approaches, the crabs return to the wave beach to burrow, but do not necessarily return to the level at which they burrowed the previous day. In B the horizontal bars show the approximate distribution of burrows on the wave beach on alternate days. The vertical bar shows the mean position and the dotted line shows variation in the extent of tidal cover. [From Baker (1978a) after Barras] although many shore-crabs may return each day to roughly the same part of the shore (e.g. Fig. 3.2), they do not return to precisely the same spot. Yet others (e.g. Patas Monkey, Fig. 3.1) show no vestige of a daily return migration and migrate in a more or less straight line from one night's roosting site to the next. Among humans, only some hunter–gatherers (e.g. desert-dwelling Australian aborigines) and some pastoral nomads (e.g. Bedouin) regularly establish sleeping sites that are abandoned after only one or a few nights.

Animals that change their sleeping site every day can still be described as having a *daily migration cycle*, even though it is not a *return* migration cycle. For all animals, therefore, we can usefully imagine the daily migration cycle as being a building block for the lifetime track, whether or not it is a return migration cycle.

If the daily migration cycle is the building block of the lifetime track, then the gross form of the track depends on the way that these blocks fit together. Without forgetting that the study of migration is concerned with understanding *tracks*, we can make life easier for ourselves here by distinguishing between an individual's *track* and its *range*. A day's track is the path traced out by an animal's body during the course of a day. A day's range, on the other hand, is made-up of all the sites visited by an animal during the course of the day. The range is simpler than the track because no matter how many times the track passes through a site in the course of a day, it only appears on the range once (Fig. 3.3). As well as daily tracks and ranges, we can also refer to monthly, seasonal, annual and, of course, lifetime tracks and ranges.

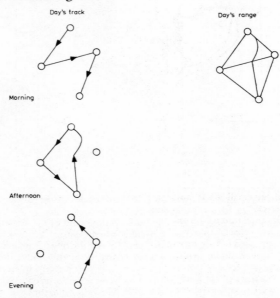

Fig. 3.3 The distinction between track and range
The day's range is obtained by superimposing the day's tracks.

Even those animals, such as most humans, that have a daily return migration cycle rarely have daily ranges that are identical from one day to the next (e.g. Baboon, Fig. 3.1). Nevertheless, *degree of overlap* of successive day's ranges is usually greater than for animals that change their roosting site every day (e.g. Patas Monkey, Fig. 3.1). Animals, such as the Patas Monkey, that show minimal overlap between successive day's ranges are often referred to as *nomadic* (i.e. without a fixed home).

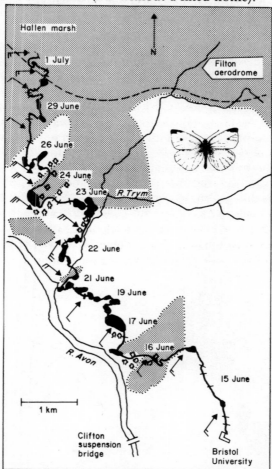

Fig. 3.4 Track of the small white butterfly, *Pieris rapae*: an example of the linear range form of lifetime track
⊢⊣, track of an individual butterfly from first sighting to the place at which it was lost from view. Upon such loss, observation was maintained at the place of disappearance until a second individual appeared following the same track as the previous individual. Solid black areas along the track show sites in which flight ceased to be straight and in which courting, oviposition, roosting and feeding occurred. Unshaded areas are urban, with some gardens. Stippled areas are expanses of flat open grassland. Wind speed and direction are shown by arrows: one short feather = one unit; one long feather = two units on the Beaufort scale. The dashed line indicates the boundary of the city and old county of Bristol.
[From Baker (1978a)]

Inevitably, a month's range is greater than any of the day's ranges from which it is built. It does not follow, however, that the month's range is greater for nomadic animals than for similar animals with a fixed roost (compare the range size of the Baboon and Patas Monkey in Fig. 3.1). The size and shape of a month's range for nomadic animals depends on their track pattern. Many, such as the Patas Monkey and human hunter − gatherers, although nomadic, stay within a *limited area range*. Others, such as some insects, straighten out their track and produce a *linear range* (Fig. 3.4).

Individuals of many species show very little shift in their monthly range, their year's range being more or less the same as each of the monthly ranges from which it is built. Industrialist humans show this pattern; so, too, do agriculturalists. Occasionally, such humans do migrate to a completely new range, but they do so irregularly and usually at intervals greater than a year. By way of contrast, some individuals of other species are continually shifting their monthly range (e.g. the female Bank Vole in Fig. 3.5). More often than not, however, if the monthly range shifts at all, it does so on a seasonal basis, the individual using one range for the duration of one season, then shifting to a different range for the duration of the next season. Sometimes this shift in range is simply an expansion and contraction, one season's range being larger than, but containing, another season's range (Fig. 3.6). In the best-known type of seasonal shift, however, different season's ranges are completely separate and the individual moves between them by means of a seasonal migration.

Fig. 3.5 Shift in home range with time: the history of a female Bank Vole, *Clethrionomys glareolus*, on Skomer Island, Wales
[From Baker (1978a) after Jewell]

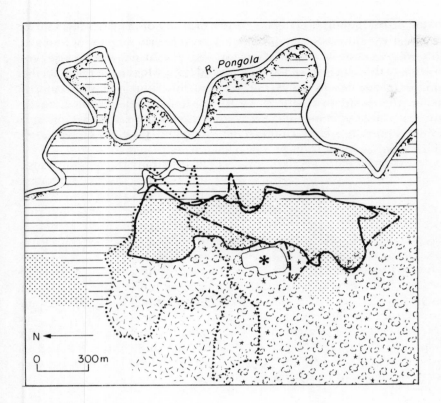

Gallery forest

Floodplain grassland with isolated large trees

Floodplain margin community dominated by *Acacia* thickets

Thicket vegetation

Open woodland savanna

Game ranger camp

Fig. 3.6 Seasonal change in size and degree of overlap of home range: the Vervet Monkey, *Cercopithecus aethiops*
Dotted line, year's home range of group A; Solid line, dry season home range of group B; dashed line, wet season home range for group B.
[From Baker (1978a), after Moor and Steffens]

When an individual lives longer than a year, seasonal migrations lead to an annual migration cycle in the same way that hourly migrations lead to a daily migration cycle. More often than not, annual migration cycles are return migration cycles, ending in a season's range that is more or less the same as the one occupied a year previously. If there are only two seasonal ranges, the result may be a *to-and-fro* migration cycle. Some animals, however, such as Mountain Sheep, some human pastoral nomads, and various seabirds may have a number of 'seasonal' ranges in the annual cycle, which then takes the form of a *migration circuit* (Fig. 3.7).

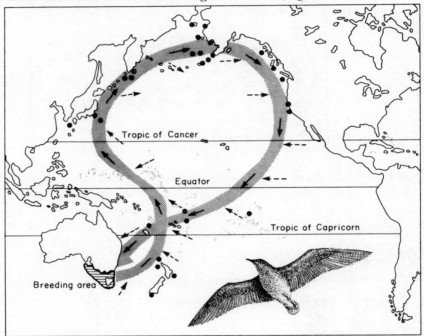

Fig. 3.7 Migration circuit of the Slender-Billed Shearwater (= Tasmanian Mutton Bird), *Puffinus tenuirostris*
Arrows on the stippled loop show the direction of migration. Dashed arrows indicate prevailing wind direction. Solid dots show the positions of collected specimens.
[From Baker (1978a), after Serventy (from Dorst and Orr)]

Fig. 3.8 Re-migration circuits of the Desert Locust, *Schistocerca gregaria*
The maps show the seasonal distribution of Desert Locust swarms as determined during the period from May 1954 to May 1955. Arrows show the migration directions in the different areas and in the bottom right-hand map these arrows are collated to demonstrate major migration circuits. These are quasi-regular, though precise routes and destinations show year to year variation. The seasonal position of the Inter-Tropical Convergence Zone (ITCZ), a major rainfall area, is indicated by a heavy line for the months of January and July. Four major demes may be recognised, each with its own migration circuit: (1) the eastern Arabia, Iraq, Iran, Afghanistan, Pakistan, northern India deme; (2) the western Arabia, central and eastern North Africa deme; (3) the West and North West Africa deme; and (4) the East Africa and southwestern Arabia deme.
[From Baker (1978a), after Rainey and Aspliden]

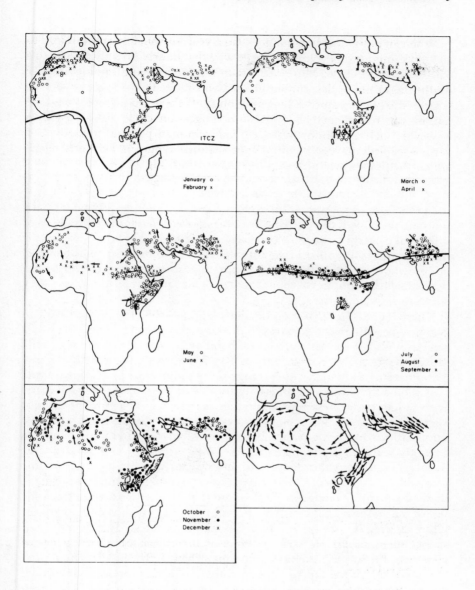

When an individual lives less than a year, annual migration cycles are completed not by the animals that performed the first leg of the cycle, but by their offspring or their offspring's offspring. The Desert Locusts that complete a vast annual circuit around the edge of the Sahara Desert (Fig. 3.8) span two or more generations, each generation carrying on where the parental generation left off—an entomological version of a sprint-relay team of athletes but spawning offspring instead of handing over a baton. Such migration cycles are not return migration cycles as described above because the individuals that return to the starting place are not the ones which originally set out from it. In order to distinguish the two, the relay-type of cycle is termed a *re-migration* cycle.

Not all long-term migration cycles are annual. The classic exponent of a non-annual cycle is the Atlantic Eel, *Anguilla anguilla*, some individuals of which take nearly 20 years to complete their circuit. These eels are also the prime example of animals that complete just one circuit in their lifetime; lampreys are another (Fig. 3.9). Such migration cycles are *ontogenetic*, individuals returning to reproduce and die at the place where they themselves were born.

There is one other major ontogenetic aspect of the lifetime track. When young, a vertebrate not only travels through a larger range than when adult but also visits a wider variety of habitats, many of which are quite unsuitable for the species concerned. To a large extent, the same is also true for invertebrates, whether their range is of the linear or limited-area type.

It is no accident that most descriptions of lifetime tracks are concerned with ranges rather than with the tracks themselves. This is because, until very recently, most of the methods available for studying migration gave information concerning ranges, but not tracks. In the past, the approach has been to mark an individual in some way, attempt to recapture it at intervals, and from these events to attempt to build up a picture of the individual's movements. Many ingenious marking methods have been devised and are reviewed in Stonehouse (1978). Methods range from the use of natural markings, such as the bill patterns of swans or the faces of Chimpanzees, to spraying butterflies with harmless dye; painting or scratching numbers on snails, crabs and tortoises; clipping the scales of reptiles or the toes of amphibians and mammals; attaching disks or clips to the fins of fish or the ears of large mammals; placing numbered or colour-coded rings, bands, or tags on the legs or wings of birds or bats; firing stainless-steel pins into whales; and introducing by feeding or injection a harmless radioactive chemical that appears over an extended period in the animal's faeces.

An alternative to marking individuals is to observe demes or populations of species and try to deduce the movements of individuals from the

Fig. 3.9 Life history and ontogenetic return migration of the anadromous River Lamprey, *Lampetra fluviatilis*.
[From Baker (1978a)]

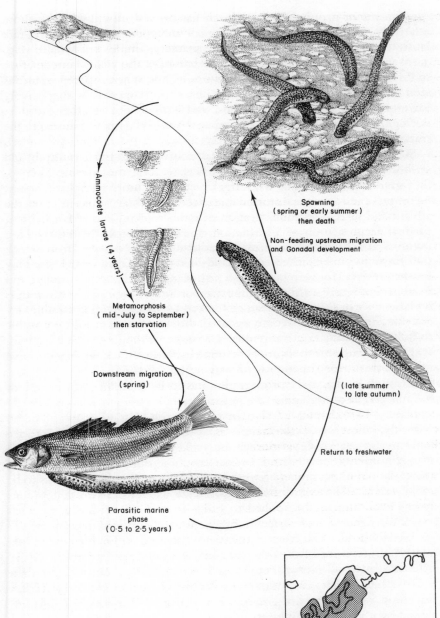

Ammocoete larvae (4 years)

Metamorphosis
(mid-July to September)
then starvation

Spawning
(spring or early summer)
then death

Non-feeding upstream migration
and Gonadal development

Downstream migration
(spring)

(late summer
to late autumn)

Return to freshwater

Parasitic marine
phase
(0·5 to 2·5 years)

movements of groups. This approach has proved most useful for those relatively few vertebrates for which all members of a deme migrate latitudinally. Temperate birds are the prime example, but whales (Fig. 3.10) and, to some extent, pinnipeds and bats have also yielded information to this approach: the 'analysis of sightings'. Thus, when a species of bird is seen at one latitude during the breeding season, appears at successively lower latitudes as autumn proceeds, and is then seen at other latitudes during the winter, the track of an 'average' individual may be similar to the track of the deme.

Such techniques have been invaluable in lifetime-track research and without them information would have been scarce in the extreme. Even so, they give a picture of ranges, not tracks; a picture, moreover, that is at best incomplete and at worst distorted due to recapture bias. In recent years, the advent and increasing sophistication of radio-tracking has at last led to a gradual accumulation of information concerning tracks. Radio-tracking involves attaching to the animal a radio-signalling device, from which transmissions can be picked up by hand-held or fixed receivers. There may be one receiver (for searching), two (for triangulation), or a complete and fixed grid of receivers (for continuous recording). Some large transmitters on large deer or Polar Bears can be tracked by Earth-orbiting satellite. By attaching transmitters that emit signals of different frequencies, it is possible to follow more than one individual at a time. Although we are still a long way from being able to record the complete lifetime track for an individual of any animal, the situation is improving all the time.

Meanwhile, the study of migration has to be brought into the age of behavioural ecology using the information on ranges that is already available. This chapter has shown that by thinking in terms of daily, monthly, and seasonal (etc.) ranges as building blocks, it is possible to see patterns and relationships between the tracks of animals from a wide range of taxonomic groups. From these patterns, questions begin to emerge. If the variation in the human lifetime track is as comparable as it seems to be to the variation shown by other vertebrates, to what extent can conclusions reached for humans be applied to other animals? Conversely, to what extent can conclusions reached for other animals be applied to humans? We are used to the idea that when humans move around within their individual yearly ranges, most of the time they know where they are going. Many other animals look extremely purposive as they move around within their yearly ranges. How many of these also know where they are going? If they do know where they are going, when and how did they first become familiar with the different locations within their range? How do they, time after time, find their way back to these locations? Why do they then sometimes abandon one range and migrate to another? Why do species and individuals differ in their pattern of shift of daily, monthly and seasonal ranges?

These are the important questions. They emerge naturally from the new

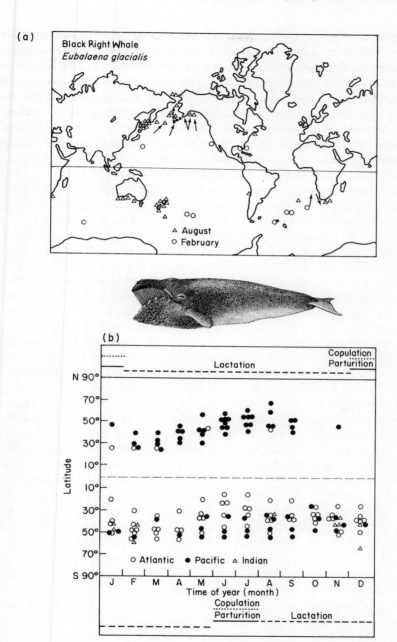

Fig. 3.10 Seasonal return migration of the Black Right Whale, *Eubalaena glacialis*, as indicated by an analysis of sightings

Each dot shows a sighting of a Right Whale. Arrows in (a) show possible migration 'legs'. There is a clear polar migration in summer and a tropical migration in winter. There is also a clear separation of the demes of the northern and southern hemispheres.
[From Baker (1978a)]

regime's view of animals and their behaviour; not least from the view that human behaviour is as much a part of the raw material for study as that of any other animal. Perhaps the most important of these questions is that of familiarity. Any animal that, like Man, has a strong *sense of location* derived from social communication, exploration and learning is clearly the absolute antithesis of the ethological concept of an animal but the perfect manifestation of the behavioural ecology concept. It is imperative, therefore, that before going any further we identify what range of animals fall into this category. The chapter that follows is concerned with just this problem.

4

A sense of location

For decades students have been taught that the cardinal sin for a behaviourist was to think anthropomorphically: to bestow upon other animals those thoughts, feelings and emotions that are unique to the human condition. The belief that humans are unique (i.e. superior) in all things cerebral and emotional is deep-rooted. It was inevitable that when the new regime of behavioural ecology, particularly sociobiology, began to challenge this belief there would be a strong reaction. Much has been written about 'the sociobiology controversy' and doubtless more is to follow (for balanced biological, sociological, and philosophical reviews respectively see Parker 1978, Brownstein 1978, and Ruse 1979). Where evidence is becoming available, however, the pillars of human uniqueness such as genetic make-up and language (see review by Desmond 1979) are crumbling. So far, migration has not entered the fray. It is time for it to do so, for it has a lot to offer.

The anthropomorphic view of migration would be that Man is not the only animal with a *sense of location*; that other animals also know where they are and where they are going. Moreover, they do so in no less cerebral a sense than Man. In this and the next few chapters I shall argue that this anthropomorphic view is, after all, the correct one, that studies of the lifetime track of other animals are extremely relevant to an understanding of the human lifetime track, and that, even more to the point, studies of the lifetime track of humans can tell us a great deal about the lifetime tracks of other animals.

Consider your own track during the past 24 hours. It can be broken down into a number of migration units. The majority of these units will have been aimed at particular destinations and, in general terms, each of these destinations provided a resource. Your day's range will have consisted of a number of *resource locations* and the *routes* you used to travel between them. The chances are that most of these destinations and routes you had used before and had memorised. In other words, you were *familiar* with them. The total number of all the sites and routes with which you are familiar make up your *familiar area*. In principle, your familiar area could consist of all the places you have ever visited in your lifetime. Memory being what it is, however, you can probably only claim to be familiar with a fraction of these. Moreover, you will be more familiar with some than others. Nevertheless, it is certain that during the past 24 hours you will have made use of only a small proportion of these: your daily range is much smaller than your familiar area.

When a range is part of a familiar area it is termed a *home range*. Not only was your last day's home range smaller than your familiar area but so too will be your next month's and year's home range. Indeed, it is probable that there are parts of your familiar area that you will never visit again. Presumably this is because you have no further use for them; either the cost of visiting them is too great or the resources they offer are no longer valued by you. Perhaps, even more likely, the resources offered are either no longer required or are more readily available elsewhere. In other words, these sites are *ranked* very low in terms of *suitability*.

All of the sites with which we, as individuals, are familiar are ranked according to suitability. We know not only where we can obtain all the resources we are likely to need but we also know that one place is the best for obtaining one resource but some other place is best for obtaining another. The memorised information that we possess concerning our familiar area is being used continually. Each time we need something we use this information to decide where to go. We weigh up the expected *cost* (time, energy, money, danger) of getting to the various sites that we know from past experience provide the resource. We also predict the relative suitability of the sites. We then balance these two factors (cost and suitability) for each site and *calculate* which is best. Having made our decision, we then migrate to the site concerned. Such a migration can be termed a *calculated migration*. In effect, all of the migrations within a familiar area are calculated migrations.

Organising the lifetime track so as to spend most of one's life in a familiar area is a very efficient method of exploiting the environment. The main reason for this lies in the advantage of calculated migration over its alternative, which may be termed *non-calculated migration*. Knowing where to go and more or less what to expect at the destination clearly carries much less risk and is much more efficient than forever journeying into the unknown in the hope of finding what you are searching for. The advantages of living within a familiar area are so clearly evident that it would be a very great surprise if Man were the only species to organise its lifetime track in this way. The problem is how we can tell whether other species also live within a familiar area. The reason we know that Man does so is partly because of our own subjective experience and partly because we can communicate with others of our species. Suppose, however, that you and I were an alien intelligence, visiting Earth for the first time. How could we tell that humans live within a familiar area? Moreover, having decided on how to obtain objective evidence, what range of animals should we conclude organised their lifetime track in a way similar to Man?

The main characteristic of living within a familiar area is that the individual is motivated and able to visit sites that it has visited previously in preference to sites that it has not. In consequence, when an individual over an extended period shows a high degree of return to a site, the chances are that it is doing so because it is familiar with the site; the migrations are probably calculated. It is always possible, however, that the animal ends up in the same site time after time by chance. Although a high degree of return to one or a number of resource locations may be a strong indication of a familiar area, as evidence it is not entirely unequivocal. Critical proof requires evidence that the animal uses a particular site in preference to a comparable site that it has not previously visited or has visited and rejected. The simplest way to obtain such evidence is to carry out *displacement– release*, or *homing*, experiments. If an animal that usually migrates from A to B is displaced from A to C yet still migrates to B, the result is strong evidence that the animal is both motivated and able to migrate to memorised site B rather than some other site. The evidence is particularly strong if C and B offer similar resources (e.g. are both suitable roosting sites).

The vast majority of all migrations performed by humans are directed at destinations which the individual has visited previously. When humans are forcibly displaced from their home range (e.g. prisoners of war), upon escape or release they most often return to their original home. Only if the cost of return is high by virtue of displacement distance (e.g. for slaves displaced from Africa to America) or if the place of release is judged to be more suitable than the original home (e.g. some prisoners of war) does the individual not return. There is, then, objective evidence that most of the time humans are motivated and able to visit sites with which they are familiar in preference to sites with which they are not. In other words, humans live within a familiar area.

Everyone who reads this book will be an industrialist; in their mind's eye they will relate the above paragraphs to the conventional industrialist's lifetime track of a home range that shows little, if any, seasonal shift. Few would doubt, however, that the same applies to hunter–gatherers and pastoral nomads. Whether nomadic or performing seasonal migration cycles, all humans live within a familiar area. The daily shift of hunting and gathering sites of hunter–gatherers and the seasonal shifts of range of pastoral nomads are all aimed at sites with which they are familiar. Whether an individual human lives within a fixed range throughout the year or shifts on a daily or seasonal basis depends entirely on the environment; on the stability or shifting pattern of resources. Quite possibly the familiar area of the average modern industrialist is little different in size from that of pastoral nomads or hunter–gatherers. The nature of the industrial environment is such, however, that few individuals are faced with resources that shift on a daily or seasonal basis. Nevertheless, some are. The annual track of a commercial traveller is comparable to that of a hunter–gatherer; that of a cotton harvester in the southern United States travelling north with the ripening cotton is comparable to that of a pastoral nomad; and so on.

For humans, then, an important fact emerges. All individuals establish a large familiar area. Out of this familiar area there crystallises a year's home

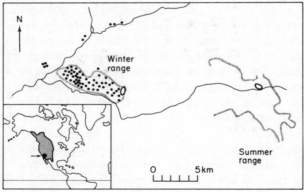

Fig. 4.1 Seasonal return migration of a herd of Mule Deer, *Odocoileus hemionus*, on the west slope of the Sierra Nevada in Tulare County, California
Females were trapped and marked with either a bell or a radio transmitter in the summer home range in July and August. The areas outlined by dashes are the main winter and summer ranges of the herd. Dots show winter sightings of deer with bells. The small area of about 1 km diameter shows the home range of a single individual in summer (after trapping) and the following winter as determined by radio-tracking. In spring, the female returned to the same home range as occupied the previous summer, accomplishing the 15 km migration in less than three days.
[From Baker (1978a), after Schneegas and Franklin. Photo by Dr. C. G. Hampson]

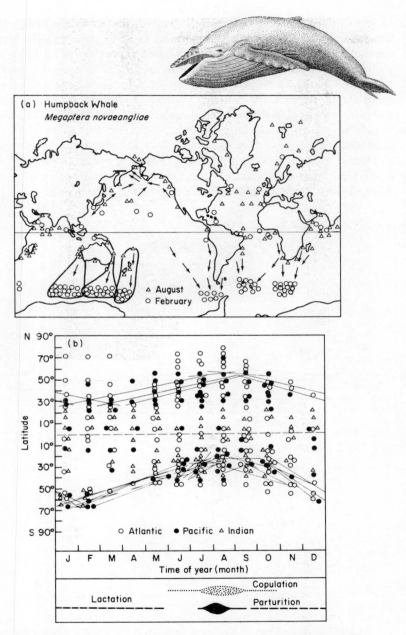

Fig. 4.2 Annual migration cycle of the Humpback Whale, *Megaptera novaeangliae*
Each dot shows a sighting of a Humpback Whale. Arrows in (a) show possible migration
legs. Lines in (b) connect places of marking and recapture of tagged individuals. Individuals
marked within each of the three areas indicated in the southern Pacific and Indian Oceans are
invariably recaptured within the same area suggesting a high degree of return to winter and
summer ranges used in previous years.

range. The form of this home range depends on the stability or pattern of shift of resources within the familiar area. In view of the human situation, we should be alert to the possibility that other species may be living within a familiar area no matter what form their annual home range may take.

All non-human terrestrial mammals seem to live within a familiar area comparable to that of humans. Where investigated, all insectivores, rodents, primates, carnivores and ungulates, etc. are found to show a high degree of return to sleeping, feeding, drinking and other sites. When individuals live longer than a year and show a seasonal shift in home range, degree of return is still high (Fig. 4.1). Homing experiments have shown that rodents and insectivores will often home from distances of 700 m or more (Bovet 1978). Grizzly Bears, *Ursus arctos*, have homed from 45 km.

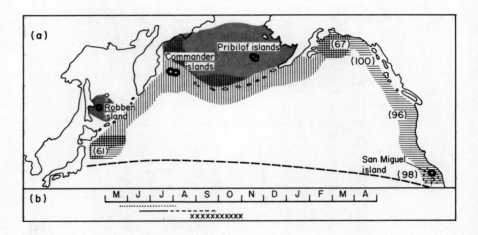

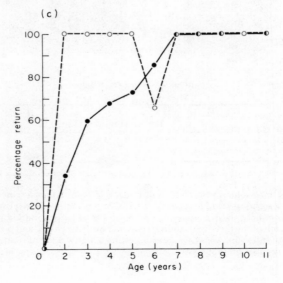

At sea, whales marked within a particular area of ocean are usually recaptured in the same area (Fig. 4.2) and pinnipeds show a high degree of return to their breeding shores (Fig. 4.3). Bats show a high degree of return to all of their season's home ranges. Homing experiments have shown that they will return, often from distances greater than 100 km, not only to breeding and hibernation home ranges but also to *transient home ranges* used as stopover and refuelling areas during seasonal migrations.

It is well known that individual birds will return to the same nesting sites, often after a seasonal return migration tens of thousands of kilometers in length. Evidence is also growing that individuals show a similarly high degree of return to winter and to transient home ranges. Homing experiments are usually carried out with respect to breeding home ranges.

Fig. 4.3 Degree of return of males and females of the Northern Fur Seal, *Callorhinus ursinus*
(a) Named islands edged by thick line, breeding areas; stippled area, feeding distribution of males (August–November), females (July–November) and immatures (September–December); vertically hatched area, feeding distribution of males (November–May); horizontally hatched area, feeding distribution of females (November–June); dashed line, approximate edge of oceanic feeding distribution of immatures; figures in brackets, percentage of individuals in area that are adult females.
(b) Seasonal behaviour. Dotted line, territorial period; solid line, parturition period; dashed line, lactation; crosses, moulting period.
(c) Degree of return to the natal shore. Percentage return refers to the percentage of individuals, born on the Pribilof Islands, that, when they return to the Pribilof Islands, return to the shore on which they were born. Open circles, females; solid circles, males. Females usually produce their first young when 5 years old, or even later. Males begin to copulate when 8 years old but are not successful at obtaining a territory and harem until about 12 years old.
[Compiled from Baker (1978a), after Kenyon and Wilke, King, Chugunkov and Prokhorov, and Peterson, LeBoeuf, and DeLong. Photo courtesy of National Marine Fisheries Service of the United States Department of Commerce]

Homing is good whether the individuals shows no seasonal shift of home range (e.g. Homing Pigeons) or whether the breeding home range is just one part of an annual migration circuit (e.g. Manx Shearwater, *Puffinus puffinus*; Fig. 4.4).

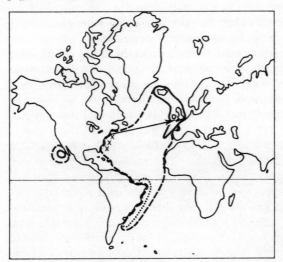

Return of young to breeding grounds in northern summer

Age

1   None (x)
2   Few (June / July)
3   May / June
4   April
5   Breed

Fig. 4.4 The migrations and homing of the Manx Shearwater, *Puffinus puffinus*
The continuous line surrounds the breeding islands of two demes of the Manx Shearwater. The Atlantic deme spends from September to February off the east coast of South America (dotted line). Young birds do not return to the breeding grounds during their first independent summer but instead wander over large areas of the North Atlantic, visiting the Atlantic coast of North America (x). The arrow shows one well-known example of homing following experimental displacement.
[From Baker (1978a), after Matthews, Orr and Mead]

As adults lizards and snakes live within a limited area and return time after time to particular basking, feeding and hibernation sites. Homing experiments on small lizards show that they return to their previous range from distances of up to 100 m. When it rains on the Galapagos Islands, Giant Tortoises, *Testudo gigantea*, converge on mud-holes which they seem to have learned collect and retain water. Painted Turtles (= Terrapins), *Chrysemys picta*, in the United States return to the same stretch of water after journeys of several kilometres. So, too, do Spectacled Caiman, *Caiman crocodilus*, in Venezuelan freshwater lagoon areas. Homing experiments on the latter species show they will return to a particular freshwater lagoon even when displaced to another 1 km or so away (Gorzula 1978). At sea, female Green Turtles, *Chelonia mydas*, return to the same breeding shore every 2–3 years. Homing experiments off Cedar Keys, Florida, show that the species will return to feeding grounds from at least short distances.

Frogs, toads and salamanders return each year to spawn in the same breeding ponds or the same 50 m stretch of river. Homing experiments have shown that individuals of these amphibians will return to their breeding sites from a kilometre or more. Male frogs and toads will return to join the breeding chorus in their 'home' pond even when displaced to a nearby pond that also contains singing conspecifics.

Radio-tracking studies on White Sturgeon, *Acipenser transmontanus*, in the Mid-Columbia River, USA (Haynes *et al.* 1978) have shown that an individual may alternate between sites 10 km apart several times during the course of a year. Individual trout may use the same spawning range year after year and after spawning return downstream to the same stretch of river in which they fed the previous year. Salmon return as adults to spawn in the same tributary as that in which they themselves were spawned. If displaced to some other tributary in the same drainage system they nevertheless return to their original choice. In lakes, fish return to their individual spawning range at the lake's edge if displaced and released at the lake centre (Fig. 4.5). On the shore, individual inter-tidal fish use the same rock-pool at each low tide and return to it if displaced. Individuals of species of oceanic fish, such as Herring, *Clupea* spp, Cod, *Gadus morhua*, and Plaice, *Pleuronectes platessa*, return to the same spawning grounds year after year.

Using the criterion of repeated use of an area and the motivation and ability to return to a particular site when forcibly displaced, Man seems to share with all vertebrates the habit of living within a familiar area, no matter what form the yearly track may take. But what of invertebrates? Life within a familiar area would seem to require a fairly sophisticated spatial memory. Are invertebrates barred from living within a familiar area through some limitation of their central nervous system?

Honey Bees, *Apis mellifera*, make frequent return journeys between the hive and particular feeding sites often after travelling several kilometres between the two. If displaced while feeding on a bowl containing sugar solution, the bees can still find their way home. If the hive is displaced while a bee is foraging, the insect returns to where the hive used to be. What would a human do if his house were displaced to a nearby street while he was away? Hunting wasps that bury paralysed caterpillars in previously-dug holes in the ground as food for their future larvae can still locate the hole after displacement and subsequent release.

On the shore, Limpets, *Patella* spp, return at each low tide to rest in the same circular depression in the rock surface that they occupied 12 h previously. Homing experiments show that the depression can still be

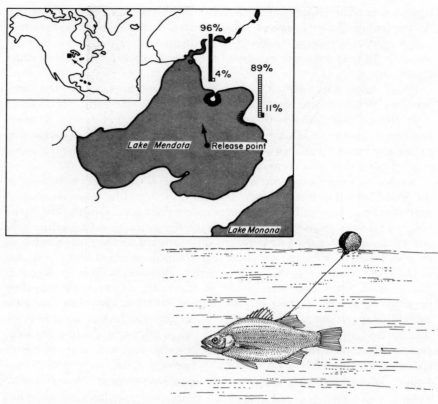

Fig. 4.5 Homing by the White Bass, *Roccus chrysops*, in Lake Mendota
The White Bass has two spawning grounds in Lake Mendota (solid and horizontal
hatching), which are 1.6 km apart. Spawning takes place in May and June, the rest of the
year being spent in open water. Histograms show the returns of males collected during the
spawning season from both spawning grounds and then released at the lake centre. Of those
that were recaptured, most had returned to their point of first capture. The arrow drawn
from the release point shows the mean bearing of the released fish one hour after release as
followed by attaching a small float to each fish with a nylon line.
[From Baker (1978a), after Hasler, Horrall, Wisby and Braemer]

located after displacement. Sandy-shore amphipods and isopods alternate
between upper-shore roosting sites and lower-shore feeding sites. When
displaced they move in a way that would re-locate the appropriate level or
zone on their home shore. In the sub-littoral zone, Spiny Lobsters, *Panulirus
argus*, return to the same crevice in which to roost each day and can still
locate their own crevice even when displaced (Fig. 4.6).

On land, slugs and snails often return each morning to the same hole in
the ground or a wall. Individuals of at least two species of snails have been
shown to return each winter to hibernate in the same area as previously.
Edible Snails, *Helix pomatia*, seem able to recognise the direction of their
hibernation range from displacement distances of up to 40 m. Wolf Spiders
that live along the edges of streams and rivers can, when placed on the

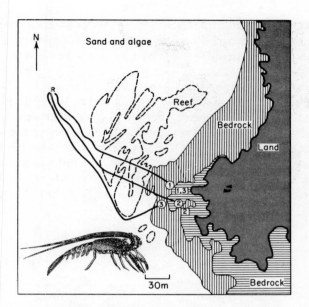

Fig. 4.6 Homing tracks of three Spiny Lobsters, *Panulirus argus*, in Great Lameshur Bay, Virgin Islands
Three lobsters (1, 2, 3) were each captured in a den (rectangle), labelled with a sonic tag, released separately at R, about 200 m from the capture point, and then tracked by hydrophone. Solid lines show the homing tracks.
[From Baker (1978a), after Herrnkind and McLean]

water, return to the bank, even if they cannot perceive it.

Confronted with this catalogue, an alien objective intelligence would be forced to conclude that all vertebrates, not only Man, and a wide range of invertebrates, organise their lifetime track in such a way that they spend most of their lives within a familiar area. Indeed, such a wide range of animals appear to have a sense of location that the problem becomes one of deciding which do not have such a sense. There are a number of obvious candidates, but even here the situation is not clear.

Butterflies, such as the Small White, *Pieris rapae*, that have a linear range (Fig. 3.4) spend their lives travelling across country in a more or less constant direction. They alternate periods of tortuous flight in habitats suitable for feeding, breeding, etc. with periods of straightened-out flight during which they search for the next such suitable habitat. Such butterflies are forever flying across unfamiliar country to unknown destinations. They have no need for a sense of location, requiring instead a *sense of direction* (Chapter 12). Displace such a butterfly sideways from its track and it does not home to its previous position but continues instead on a track parallel to its previous track. Locusts (Fig. 3.9) and all animals with a linear-range form of lifetime track are likely to behave in a similar way. Nevertheless, watch a Small White while it is feeding in a garden and it provides strong indications of having a temporary sense of location. So, too does a male

Small Tortoiseshell Butterfly, *Aglais urticae*. Adept at leading a rival away from a female that it is guarding, the male then gives the rival the slip and returns with precision to the female's last position, whether or not she is still there (Fig. 4.7).

Sedentary invertebrates, such as barnacles or those polychaetes with calcareous tubes, select a place to live at the end of a planktonic larval phase. During the search for such a place, the tiny larva visits and tests a succession of potential substrata, perceiving cues such as the nature of the surface and the presence and perhaps density of conspecifics. Early on in its search, the larva is very selective about where to settle. As time passes, if a highly suitable site is not found, the larva is more and more likely to settle in a less suitable site. We could say its threshold for what is a good place to settle

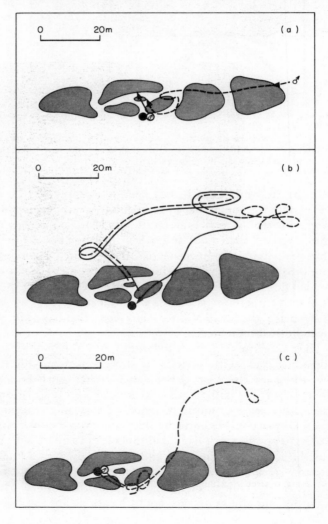

Fig. 4.7 The role of a sense of location in competitions for a female between males of the Small Tortoiseshell Butterfly, *Aglais urticae*

Diagrams show a plan view of 7 nettle patches (stippled areas). Solid dot, position of female; open dot, and solid line, position and track of original male; dashed line, track of intruding male. In (a) the intruder finds a pair by the edge of a nettle patch. In the photograph the male has landed behind the pair but usually the first male flies up and the intruder gives chase, the female remaining in position. In (b) the original male leads the intruder away from the female, often by a tortuous route, and often up to 100 m away. It then tries to disengage by sharp turns and diving. When successful, the original male returns more or less directly to the female and lands behind her. In (c), as soon as the male settles behind the female he taps her with his antennae and she immediately takes flight. The male follows and the pair fly to a new location. The intruder also returns to the original site, often only a few seconds behind the original male. A male guarding a female may have to negotiate many such encounters during the course of an afternoon. Apart from these interruptions, the pair stay in the position shown in the bottom photograph until evening when they drop into the nettles to roost. It is at this point that copulation occurs.

[From Baker (1978a). Photos by R. R. Baker]

decreases with time. If the larva fails to find a site, even of low suitability, then metamorphosis is delayed. The point of interest to us, however, is whether even a humble polychaete or barnacle larva has a sense of location while searching for a place to settle. Does it recognise a site as one that has been visited before? Having visited a site that is not suitable enough for settling early on in the search but is above threshold later on in the search and better than any discovered subsequently, can it return to that first site? Can it avoid sites visited previously and found to be highly unsuitable? I do not know the answer to these questions but intuitively would not expect a barnacle larva to show such cerebral behaviour. Instead, I should expect it to employ a pre-programmed search pattern and to settle in the first site it encounters that is above its present threshold. I referred to this expectation as intuition, but perhaps it is simply human prejudice.

Zooplankton, moving up and down in the water body that carries them (Fig. 4.8) would seem to have little use for a sense of location. A sense of direction (i.e. up and down), depth and timing would seem to be their main

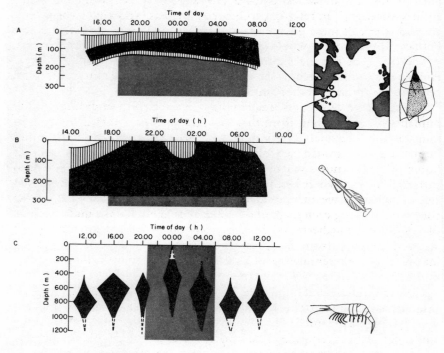

Fig. 4.8 Daily vertical migrations of three species of marine zooplankton
The stippled area shows the hours of darkness. In A and B the main concentration of animals is indicated by solid shading and lesser concentrations by hatching. The three species shown are: A, the siphonophore, *Eudoxoides spiralis*, off Bermuda; B, the chaetognath, *Sagitta bipunctata*, also off Bermuda; and C, the bathypelagic, natant decapod crustacean, *Gennadas elegans*, off New Jersey.
[From Baker (1978a), after Moore, Waterman, Nunnemacher, Chace and Clarke]

requirements. The best-known of all zooplankton, the Antarctic Krill, *Euphausia superba*, forms swarms that may stay together from hatching onwards. These swarms are carried round in ocean gyrals, reproducing when they arrive in certain parts of the waters off Antarctica (Fig. 4.9). As well as a daily vertical migration, therefore, this and many other zooplankton perform a large migration circuit. To what extent it is a re-migration, comparable to that of locusts (Fig. 3.9) and to what extent it is a return migration, comparable to that of Herring (Fig. 9.8), individuals returning to an area used previously to breed, is unknown. Unlike locusts, Herring have a strong sense of location, returning year after year to spawn in the same general area. So what of Krill: will they be found to have a sense of location?

We began this chapter by asking whether Man was the only animal with a sense of location. We end it by wondering whether there are any that do not have such a sense. Even animals with a linear range, such as many butterflies seem to use a sense of location at some times. Is this yet another pillar of human uniqueness that has crumbled? Are we justified in being anthropomorphic, in referring to other animals as 'knowing' where they are and where they are going? Many people would argue that we are not. Perhaps (the argument would go) other animals do share with Man a sense of location, but the mechanisms by which they establish and use their familiar areas are different. Man's sense of location is different from, more cerebral than, that of other animals.

The full force of this, essentially ethological, attempt to distinguish the human sense of location from that of other animals is met in relation to long-distance seasonal return migrants. When a Lapp, or some other human pastoral nomad, sets off on his journey from winter to summer home range he knows where he is going. He knows the route, where to turn right or left, where best to stop and feed or sleep, even where to go in case of inclement weather, accident or some other contingency. He knows roughly how long each stage of the journey is likely to take under a range of conditions. Furthermore, he can recall all of this in his mind's eye without moving from his present position. Armed with this spatial knowledge, this cerebral sense of location, the Man can make his journey of hundreds of kilometres with extreme efficiency.

Contrast this view of human migration with the ethological view of the migrations of salmon or birds as described in chapter 2. A salmon imprints when young on the smell of its natal stream. When, later, as a pre-spawning adult, the fish next encounters the smell of its home stream, it is triggered to swim ceaselessly against the current, to fight rapids, and to overcome all obstacles in its blind, instinctive urge to reach its natal stream and spawn. According to the clock-and-compass model, a young bird imprints on some unknown coordinates of its breeding range, migrates for a pro-grammed length of time in a programmed compass direction, and then imprints on some other coordinates of its winter range. Thereafter, as an adult,

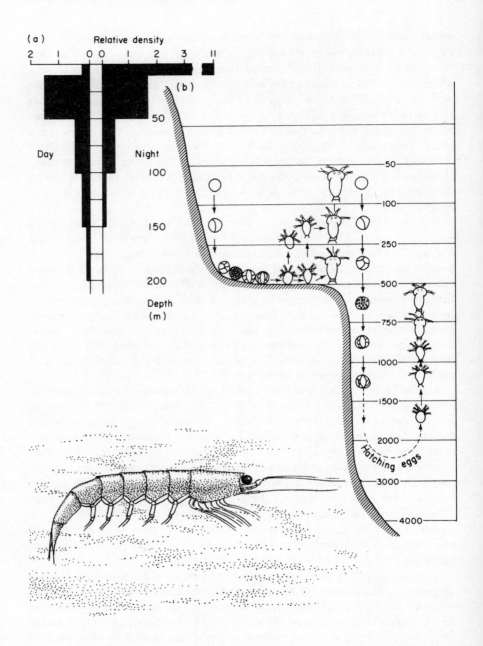

Fig. 4.9 Daily and ontogenetic return migration of the Antarctic Krill, *Euphausia superba*.
(a) Day and night vertical distribution of Krill greater than 20 mm in length.
(b) Ontogenetic return migration during early development
(c) Geographical distribution.
(d) Major surface and deep-water currents of the Southern Ocean.
[From Baker (1978a), after Marr and Slijper]

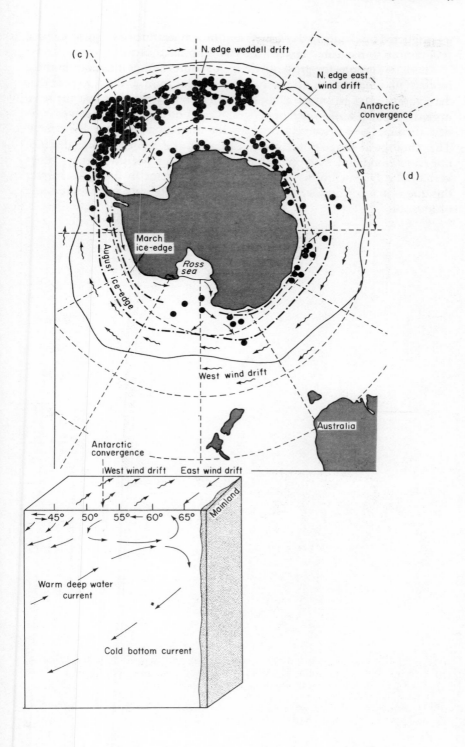

it is triggered by changing day-length and/or temperature or by some internal programme to navigate between the two imprinted sites.

There is no room for pre-knowledge of routes and destinations in these models; no judgements; no cerebral sense of location. Indeed, except that the imprints must be stored in the central nervous system and that the sense organs by which water-smell or geographical coordinates are perceived connect therein, there is scarcely need for the migration to be at all cerebral. The ethological image of the sense of location as it relates to long-distance migrants could not have been more different for Man and these other vertebrates. Do such differences exist? The first stage in trying to answer this question is to consider how the individual vertebrate achieves its sense of location.

5
Exploration: the means to a sense of location

How can we tell whether the human sense of location is fundamentally different from that of other animals? The key ought to lie with an understanding of the way that humans and other animals each build up and then use their familiar area: how they gain their sense of location.

To a human, a sense of location involves being able to picture in the mind's eye the present location relative to that of other, familiar locations. Obviously, this requires memory. Moreover it requires a special form of memory: a *spatial memory*. Nowadays, most peoples' spatial memories are aided, at least in industrial countries, by such things as maps, place-names, signposts, and roads. In terms of human evolution, however, such aids are of recent origin.

The information stored in the spatial memory is similar to, but not identical with, and certainly not as complete as, that stored on a printed map. Consider your own spatial memory. Envisage the spatial relationship between any three places that you visit regularly, such as those where you sleep, work and buy your food. The chances are that you can arrange them in your mind's eye in the form of a triangle. Perhaps, you automatically include compass direction, or at least 'up' and 'down', in the picture. The relative distances and directions between the three sites may be a little distorted or foreshortened, but they will be adequate if not accurate. However, most of the time it is not this image that you use in moving around. Whenever you wish to move between any two of these places you are most likely to use a set of self-taught instructions: turn right at the large red building; take the second turning on the left; keep on until just past the large willow tree; and so on. If the route is very familiar, it will be possible to travel without being aware of these instructions and to arrive with little, if any, recollection of having negotiated the route. In thick fog or intense dark, the usual instructions are useless. Instead you will have to switch to another set of instructions. You will have to concentrate intensely, thinking about almost every step, using detailed clues like lamp-posts, individual trees, side-roads; clues that normally you would not even notice yet are evidently stored somewhere in your spatial memory, available for recall if required.

This is the essence of the human familiar area: vague 2-dimensional maps on which all familiar locations can be arranged relative to each other and perhaps relative to some stable, absolute axis such as compass direction; detailed instructions on how to move from one place to another, such instructions being arranged in a hierarchy, so that more or less the minimum information is used for any particular migration from A to B. The sense of location merely involves knowing where on that map the individual happens to be at a particular instant in time.

Once a human is completely familiar with an area, his movements can be very efficient. Rarely does he have to travel further than necessary to obtain what he requires for the minimum cost and inconvenience. First, though, the area has to become familiar. How is this done?

Most of the people who read this book will be, or will have been, students and will probably remember vividly how it feels to arrive in a completely unknown city and to be forced to carve out from the blank terrain a familiar area of the type just described. The process is straightforward but time-consuming. Being human, as much information as possible is sought from people who have already established a familiar area. Views on the best places to eat, drink and sleep are sought, as also are the best routes to travel. Many much places and routes may be travelled in the company of other, more experienced, individuals. Sooner or later, however, most people begin a phase of independent *exploration*. This, the most important of all mechanisms in the process of familiarisation, consists of three parts: (1) *exploratory migration* (i.e. the act of moving along unfamiliar routes to unfamiliar destinations but always with the intention of eventually returning to a familiar site); (2) *habitat assessment* (i.e. judging the suitability of a route or place, identifying what resources it can offer, and ranking it against other places already discovered); and (3) *navigation* (i.e. negotiating the way back to a familiar site).

The function of exploration is simple: to visit unknown places to see what they are like. By its very nature, however, exploration is an inefficient process compared to the calculated migrations within the familiar area. Many places that are not worth visiting are visited because they are unfamiliar. Routes taken are not the most direct or convenient. Moreover, there is always the danger of becoming temporarily lost and having to search around for information to tell you where you are on the map that is being built up in the spatial memory. Some places visited during exploration are immediately incorporated within the familiar area. Others are visited only once and then avoided, ignored, or even forgotten. Yet others are *revisited for reassessment* several times before being adopted or dropped.

Moving to a new city brings the whole process of familiarisation into sharp focus because it is condensed into so short a time. There is a more insidious, longer-term process of familiarisation going on of which movement to a new place to live is just a part. When you were born you had no familiar area. Now you have. Your familiar area, moreover, probably covers thousands of square kilometres. The process by which you built up this familiar area is an exact replica of the process just described, albeit on a larger scale and spread over a longer period. If there is a difference, it concerns how conscious you are of the process. When you move to a new city, familiarisation is an urgent necessity and, at least initially, a conscious effort is made to explore, to investigate new places and find out what they are like. Lifetime familiarisation, however, is less urgent. Most people recognise the urge to travel, to see new places, but few would consciously rationalise their behaviour in terms of building up a familiar area. Nevertheless, that is what it is. Every time, given a choice, you travel an unfamiliar route to an unfamiliar destination *in preference* to a familiar route or destination, you are indicating that your motivation to explore outweighs all other motivations. Information gained during this exploration will be incorporated in your spatial memory and may be used some time in the future. Consciously or unconsciously you will have extended your familiar area.

As a child, most of your familiarisation will have taken place in the company of parents, other adults, and other children. Your initial familiar area will be a sample of the familiar area of all the people with whom you travel and associate. Independent exploration over any distance will probably not have started until early adolescence but will have gained momentum over the years.

If this picture of the way humans build up and use their familiar area is correct, we are now in a position to answer the question posed at the end of the last chapter: are the senses of location of Man and other animals fundamentally different? The key features that emerge for humans and could distinguish them from other animals are: (1) exploration; (2) a spatial memory; and (3) habitat ranking. All of these features are cerebral and all are absent from the ethological models of the long-distance migrations of birds and fish and, as we shall see, from the models of the short-distance 'dispersive' movements of these and other animals. Exploration requires cerebral judgements to be made. New sites have to be assessed and then ranked against other sites that cannot be compared directly because they are out of present perceptual range. Habitat ranking involves comparing a site that can be perceived with a site that cannot. Information has to be retrieved from the memory and compared with current information. Then, later, when a decision has to be made about which site to visit, sites have to be compared, none of which can be perceived. Exploration involves the solution of spatial problems: how to find the way back to the pre-exploration familiar area; how the position of the new site relates to previously familiar sites. As humans we take such cerebral activity for granted, but many doubt that other animals are capable of going through similar thought processes; hence the urge to distinguish the human sense of location from that of other animals.

Now we have what seems to be the key. If other animals build up their familiar area by exploration, if they rank habitats according to suitability, and if they show some sign of storing this information in a spatial memory, there can be no possible grounds for maintaining that the human sense of location is in any way unique. Investigating the existence and workings of the spatial memory takes us into the realms of navigation and is discussed in Chapter 10. The next chapter is concerned with which animals, if any, share with Man the behaviour of exploration and habitat ranking and thus have a cerebral sense of location of the human type.

6

Recognising exploration

How can exploration be recognised under natural conditions? An individual travelling through an area that it has never visited previously must be involved. Usually return, often frequent return, to a familiar location is involved. Visits to a wider range of sites than normal, poor as well as good are always involved. Only when an area has been explored and the different sites assessed and ranked can the poor ones be avoided.

The antithesis of animals that explore was described in Chapter 4 in relation to barnacle larvae. It is an animal without a sense of location; without a spatial memory; an animal with a threshold for settling that does so as soon as it encounters a habitat which exceeds that threshold. Until then it continues on a fixed search path, the twists and turns of which are programmed internally, but increase or decrease in frequency (in accordance with optimal foraging strategy: see review by Krebs (1978)) in a way that depends on the nature and suitability of the habitats encountered. This type of movement, a search pattern that involves no sense of location, I have termed *non-calculated removal (NCR) migration* (Baker 1978a). When applied, for example, to the post-weaning movements of mammals and the post-fledging movements of young birds, it encapsulates the main elements of what hitherto has been called 'dispersal'. The semantic corruptions of the word dispersal have already been described and I shudder at the prospect of propagating its misuse still further. I shall do so, however, though only for the next few paragraphs, for the sake of making clear to which of the earlier models I am referring.

Among vertebrates, the concept of 'dispersal' has been applied with most vigour to mammals and birds. Do young mammals 'disperse' after weaning or do they explore? Do young birds 'disperse' after fledging or do they explore? There are pitfalls to be avoided when trying to answer these questions. 'Dispersal' and exploration have many features in common as far as the field observer is concerned. In particular, as both involve the animal travelling through an unfamiliar area then both should be characterised by a tendency to visit a wide range of sites, bad as well as good, compared with later when a suitable habitat has been found. The crucial difference between the two is that, as it travels, the explorer is continually making comparisons between present and previous locations, rejecting some, ranking others. A 'disperser', on the other hand, responds only to the habitat of the moment and settles as soon as one is encountered that is above threshold. Any evidence that a species more or less always settles in or exploits the last site visited during the movement in question favours the 'dispersal' model.

Evidence that an animal, having visited a succession of sites, returns to settle or exploit one of the sites visited earlier in the movement favours the exploration model; particularly if there is reason to suppose the site chosen to be the best of all those visited or the sites not chosen to be in some way inferior. Another feature of exploration rather than 'dispersal' would be the use of a home base from which to-and-fro forays occur, followed some time later by removal migration to a site encountered during these forays.

Armed with this means of distinguishing between exploration and 'dispersal', there is little difficulty in deciding the true nature of the post-weaning and post-fledging movements of mammals and birds; as long, that is, as we avoid the dangers brought about by lack of data. When a young small mammal is marked before leaving its parental home range, it is rare for it to be seen again until it has settled down in a range of its own. Similarly, a young bird that is ringed in the nest is not normally seen again after fledging until it is picked up, usually dead, at some distant spot. In both cases the only data available are the starting point and destination. The *impression* given is one of 'dispersal', the animals having 'settled' at the last point visited. It is not surprising that these movements have been stamped as 'dispersal', but without knowing the track pattern between the beginning and end sites such a judgement cannot be made. Whenever adequate data are available, the impression of 'dispersal' evaporates .

The best description of post-weaning migration is probably that of Myers and Poole (1961) for the Rabbit, *Oryctolagus cuniculus*. Working in Australia, these authors observed groups of Rabbits in enclosures about 0.8 ha in area. They found that movements beyond the limits of the familiar area were behaviourally different from the usual movements within the parental home range. Such longer distance movements were carried out quickly, at a steady lope, with frequent stops to investigate the new area. Each stop was accompanied by sniffing and intense observation, often taking the form of standing against objects and craning in all directions. Such forays usually began late in the evening, after the main period of feeding, and were invariably brief, only 15–20minutes or so each night, the Rabbit concerned then returning to its previous home range. Despite such movements, many young Rabbits eventually opted to settle in or near to the parental home range. Others moved away to settle but always to a destination located previously during evening forays. Throughout most of the experimental period, adjacent enclosures were separate. Occasionally, however, access from one enclosure to another was made possible. Whenever this happened, Rabbits moved from the more-dense to the less-dense deme. On one occasion, 34 Rabbits, 30 of which were immature, migrated from a deme of 115 individuals to a deme of 52 individuals, giving demes of 81 and 86. On another occasion, 6 migrated from a deme of 53 individuals to a deme of 39, giving demes of 47 and 45.

The key features of post-weaning movement by these Rabbits are thus: (1) investigatory forays beyond the previously familiar area; (2) eventual

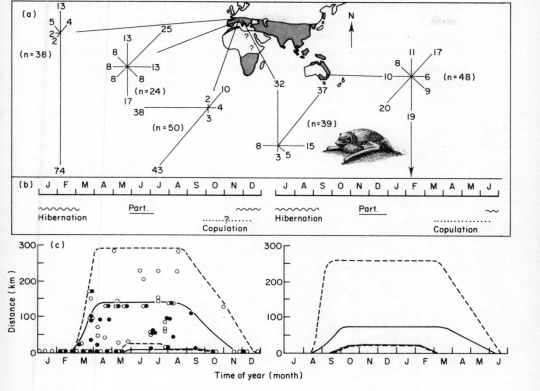

Fig. 6.1 Seasonal return migration of Schreiber's Bat, *Miniopterus schreibersii*
(a) Distribution and direction ratio. Direction ratios are shown for groups hibernating in five different study areas as shown by arrows (i.e. Czechoslovakia, north-east Spain, Germany and Austria in Europe and north-east New South Wales in Australia). Length of line and number indicate the percentage of each group that was captured in each of eight directions from the hibernation sites.
(b) Seasonal behaviour in Europe and Australia.
(c) Migration distance in Europe and Australia. A solid dot (male) or open dot (female) shows the distance from its hibernation site that a marked individual has been captured. The maximum and minimum distance limits are indicated for males (solid lines) and females (dashed lines). The figure for Australia has been compiled from a description of the migration rather than from tabulated mark−release−recapture data.
[From Baker (1978a), after Dwyer and many others]

settlement in an area visited during such forays or a decision to remain in the parental range; and (3) a clear indication that the decision where to settle is influenced by relative deme density. Such behaviour could not be more clearly exploration nor further from the concept of post-weaning 'dispersal'.

Critical data on post-weaning movements in other mammals are scarce because of the difficulty of following lifetime tracks. In a recent study in Scotland (Jenkins 1978) a young male Otter, *Lutra lutra*, was tracked by the radioactive droppings ('spraints') it left behind. The study showed that over the course of a few winter months, the Otter gradually extended its range from the lochs on which it was born to include a large stretch of a nearby river. Return to the natal lochs was frequent. The following summer the Otter moved out onto the river and eventually was using sites 70 km apart. Again the impression is one of the gradual building up of a familiar area by exploration.

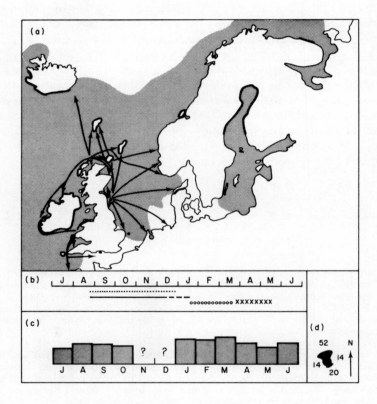

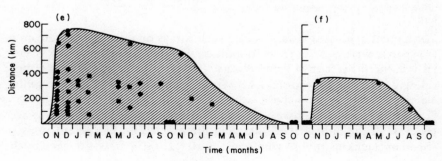

The most thorough study of post-weaning migration by bats, that of Dwyer (1966) on Schreiber's Bat, *Miniopterus schreibersii,* is also totally consistent with wide-ranging exploration. In east continental Australia, parturition occurs at the beginning of December (Fig. 6.1). Juvenile bats leave the maternity colony in late summer and spend the autumn on the move from one transient roost to another. Within a month they can be up to 150 km from the maternity roost in which they were born, the fastest rate recorded being 216 km in six days. During this movement the young bats seem to follow major linear features such as rivers or ranges of hills or mountains. In late autumn, the young bats settle down to overwinter in a cave roost along with adults. Up to this point, the behaviour could easily be interpreted along classical lines as an 'explosive post-weaning dispersal'. When, 3 months later, the bats, now called yearlings, leave the winter roost, this impression evaporates. During their spring movements the bats again travel across country, visiting a succession of roosting sites. In particular, however, many (both male and female) pay a visit to the site of the maternity roost in which they were born. After this visit, which ends before the adult females start to give birth, the yearlings spend their late spring and summer visiting and re-visiting roosts. As they do so they confine their visits more and more to the better roosts—those being used by

Fig. 6.2 Seasonal and ontogenetic return migration of the Grey Seal, *Halichoerus grypus,* in the east Atlantic
(a) Coasts edged in black, breeding distribution; stippled area, feeding distribution; arrows, some results of mark—release—recapture experiments on individuals marked soon after birth.
(b) Seasonal behaviour. Dotted line, territorial period; solid line, main parturition period; dashed line, lactation; open circles, female moulting period; crosses, male moulting period.
(c) Relative numbers on the Farne Islands, eastern Britain, for each month of the year.
(d) Direction ratio of young seals born on the Farne Islands as measured from mark—release—recapture. Figure shows percentage moving in each of four directions.
(e) Ontogenetic return migration during first two years of life. Dots, recaptured individuals; hatched area, distances from the birth site within which individuals are likely to be found.
(f) As for (e) but for adult *H. grypus.*
[From Baker (1978a), after Davies, Hickling, King, Cameron, Curry-Lindahl, and Boyd and Campbell. Photo by C. Stephen Robbins]

adults. During the autumn, some of the yearling females may be mated by males that use the autumn transient roosts as copulation sites. The following spring the majority of the now mature females, individuals that nearly two years earlier performed a supposedly explosive 'dispersal', return to give birth in the maternity roost in which they themselves were born. The behaviour is typical exploration.

The neotropical fruit bat *Artibeus jamaicensis* on Barro Colorado Island has a preference for figs and a phobia about foraging for any length of time while the Moon is above the horizon, even if it is cloudy (Morrison 1978). Usually the bats fly from their day-roost to a familiar fruit tree up to 1 km away by a direct flight lasting 5–10 minutes. New fruit trees are located by prolonged migrations of 20–45 minutes, which can only be explorations. Such explorations also only occur while the Moon is beneath the horizon.

After weaning, most pinnipeds (seals, fur seals, sea lions, walrus) begin a phase of movement that is usually described as 'dispersal' or 'nomadism' and lasts for two or more years until they attempt to reproduce. A typical example is the Grey Seal, *Halichoerus grypus,* (Fig. 6.2). After weaning, the young seals seem to spend two years travelling through an area of ocean within several hundred kilometres of the shore of their birth. During this time they are likely to haul out on distant shores, including some not usual for adults of their species. At the end of this time, however, they are most likely to return to breed on the islands, and often the shores, on which they themselves were born. Again the behaviour is much more consistent with exploration than any concept of 'dispersal'.

Ringing studies rarely produce critical evidence concerning the nature of post-fledging movements of birds, for the reasons given earlier. One of the most detailed of such studies (Mead 1979, Mead and Harrison 1979a, b), however, and the first to be written up since the first formal presentation of the exploration model (Baker 1978a, b), adopted this model in more or less its entirety to interpret the movements of the Sand Martin/Bank Swallow, *Riparia riparia*. The picture that emerged was clear (Fig. 6.3). After fledging, young birds spend several weeks before starting to move south on autumn migration travelling through a wide area around their natal colony. While doing so, the birds visit and roost in the burrows of a large number of breeding colonies. The following spring, 55 per cent of surviving juveniles return to breed in the immediate vicinity of their natal colony. As 87 per cent settle within 10 km and only 2 per cent settle more than 100 km away from their natal colony, it seems likely that most (Mead 1979), and probably all (Baker 1978a), settle to breed within the familiar area established the previous autumn. Again the behaviour is pure exploration; there is no hint of post-fledging 'dispersal'.

Only a few examples have been presented. I would claim, however, that wherever critical data are available, post-weaning and post-fledging movement by mammals and birds can be seen to be exploration and not 'dispersal'.

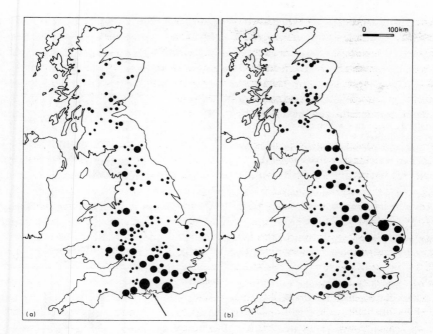

Fig. 6.3 Post-fledging migrations of Sand Martins/Bank Swallows, *Riparia riparia*

Arrows indicate the position of two roosts: Chichester (a), and Fenland (b). Dots show colonies visited by two types of bird during post-fledging migration: those ringed as juveniles at the roost, and those ringed elsewhere and captured the same autumn at the roost. Size of dot reflects number of individuals caught while visiting the colony indicated by the dot. Two major points emerge: (1) young birds wander extensively during post-fledging migration, in all directions; and (2) birds visiting the Fenland roost are more likely to wander through eastern Britain whereas those visiting the Chichester roost are more likely to wander through western Britain and visit Ireland.

[Re-drawn, with modifications, from Mead and Harrison (1979a)]

I have persevered with the term 'dispersal' in this discussion because 'post-fledging dispersal' and 'post-weaning dispersal' have, over the years, become fixed as formal terms. I advocate that these be replaced by the terms 'post-fledging migration' and 'post-weaning migration'. Such terms make no assumptions concerning function, being purely descriptive. They can be applied without corruption to a single individual showing the behaviour as well as to groups. As far as birds and mammals are concerned these migrations are probably almost always exploratory.

While on the subject of semantics, I shall, from this point on, use NCR migration, rather than 'dispersal', to describe the alternative to exploration. This frees the term 'dispersal' to be used in what I consider to be its legitimate sense: to imply the scattering of a group, the opposite of convergence.

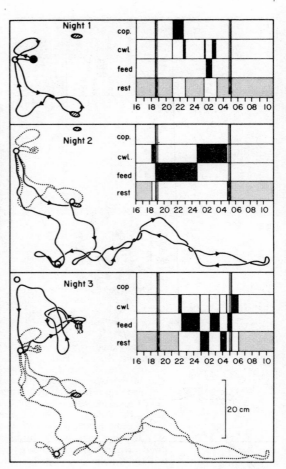

Fig. 6.4 Track pattern of a grey field Slug, *Agriolimax reticulatus*, when placed in a novel environment.

The diagrams show the movements and behaviour of an individual on five consecutive nights in an experimental arena containing topsoil and kept inside a greenhouse. The arena measured 91 × 68 cm and movements were recorded by time-lapse photography, one frame being exposed every 15 s using high-speed flash illumination. When extended, a full-grown Slug measures 3 − 4 cm.

In the diagrams, the solid line shows the track of the animal on the night in question. The dotted lines show the tracks on previous nights. Diagonally hatched areas show the positions of slices of carrot provided as food for the animal. Open circles show the positions of holes in which roosting could occur. The inset in each diagram ʒhows the incidence and timing of four different behaviour patterns: cop, copulation; cwl, crawling; feed, feeding; and rest, resting or roosting. Vertical lines indicate the times of sunrise and sunset. The x-axis shows time of day (h GMt). [From Baker (1978a), after Newell. Photo by Brian Long]

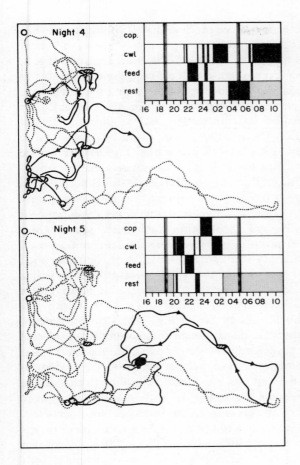

It was implicit in the interpretation of the British Sand Martin/Bank Swallow study that the birds concerned were not only familiarising themselves with the location of suitable breeding colonies but also with their suitability. This in turn implies a comparison and ranking of habitats, an important point in deciding whether birds, like humans, have a cerebral sense of location. Fortunately, we can do better than implication. When Great Tits, *Parus major,* are placed for the first time in a large aviary they soon learn by exploration to concentrate their search for hidden meal-worms in the patch containing prey at the greatest density. However, they spend more than the optimum foraging time in the less profitable patches (Krebs and Cowie 1976). When the best patch is suddenly reduced in quality, the Tits switch to foraging primarily in the second-best patch (Smith and Sweatman 1974), providing evidence that the excess visits are really RFR (revisiting for reassessment) migrations concerned with habitat ranking. This seems to show that the birds had sampled each place and stored up information about its relative profitability. Rankings were then checked and updated by RFR migrations.

As far as mammals and birds are concerned, the evidence, such as it is, seems clear: they do have a cerebral sense of location quite comparable to that of humans. The next question must be: what range of animals have such a sense? For other vertebrates, critical data are non-existent. The movements of young snakes, lizards, turtles, and amphibians are all compatible with exploration (Baker 1978a) but could be argued to be NCR migrations, even though adults all live within familiar areas. The ability of Goldfish, *Carassius auratus*, to learn to negotiate a maze has been known for many years (Churchill 1916). More recently, it has been shown that, when placed in a large (5.0 × 5.0 × 0.5 m) tank which contains only water, naive Goldfish do not show search patterns that cover the entire field (Kleerekoper *et al.* 1974). Instead, locomotion is organised spatially and temporally. The fish moves about in one specific area for a specific period after which its centre of activity shifts to another part of the tank. Such behaviour must involve a spatial memory and strongly suggests that a Goldfish, placed in a novel area, does not simply perform NCR migration but instead establishes a familiar area by a process of exploration.

Time-lapse photography of a Grey Field Slug, *Agriolimax reticulatus*, over a period of five nights after being placed in an arena (91 × 68 cm) containing topsoil and kept inside a greenhouse shows a number of features (Fig. 6.4). Directed movements to co-copulants and food are evident in Fig. 6.4 as also, at dawn or thereabouts, are directed movements back to the roosting hole used the previous night. Forays into areas not previously visited followed by a return to the original roosting hole also occurred. During such forays, food items that might be used on subsequent nights were located, as also were potential roosting holes. When removal migration to a new roosting site occured (Night 4), it was to a hole located and investigated two nights previously. The new hole seemed then to be a base from which fresh forays were made into previously unvisited areas. Similar behaviour has also been found for the snail, *Helix aspersa*, (Bailey MS). Although the trail of the slug has not been subjected to the elegant statistical analysis to which the Goldfish track was subjected by Kleerekoper *et al.* (1974), the implication is the same: ordered exploration of a novel area; investigation and assessment of located resources; use of these resources on a future occasion. The parallels with the behaviour of Rabbits described earlier are striking.

As they move, slugs and snails lay mucus trails. The dotted lines in Fig. 6.4 may have been as real to the slug as they appear in the diagram. These could have served as an aid to the animal's spatial memory but could not have replaced it entirely. The main impression from Fig. 6.4 is one of avoidance of previous trails, as might be expected if the main behaviour was exploration of the area. At other times, however, the animal seemed to follow its own trails, particularly when leaving the roost (e.g. Nights 2, 4

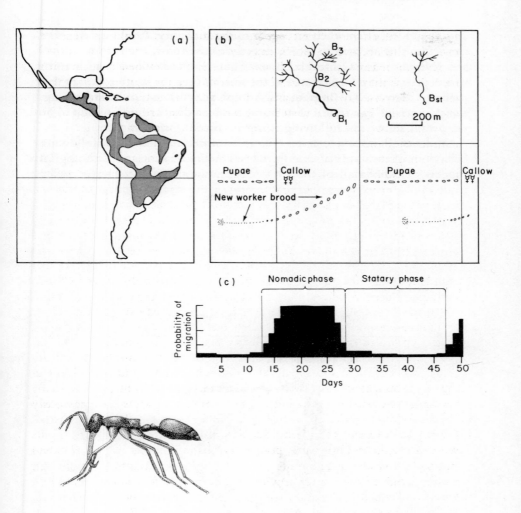

Fig. 6.5 The migration cycle of the South American Army Ant, *Eciton hamatum*
(a) The known geographical distribution of the genus *Eciton*.
(b) Raiding patterns at different phases of the brood cycle.
(c) Relative probability that bivouac-to-bivouac migration will occur on any particular night during the migration cycle.
During the nomadic phase of the migration cycle, several raiding trails radiate from the central bivouac. B_1, B_2, and B_3 mark the positions of the previous, present, and next bivouacs, respectively. Bivouac-to-bivouac migration takes place, at night, along a raiding trail of the previous day (wide line in (b)). Migration occurs nightly during the mid-nomadic phase (c). By mid-statary phase, raiding is much less intensive (B_{st} = statary bivouac), only a single raiding trail being formed and some days none at all. The brood and migration cycles form an adaptive complex. The nomadic phase begins with the emergence of a new (callow) batch of workers and ends with the spinning of cocoons by the mature larvae of the next batch. Oviposition occurs mid-way through the statary phase.
[From Baker (1978a), after Schneirla]

and 5) and less often when returning to the roost (e.g. Night 2). At other times, the slug seemed to return directly to the roost, even from a novel area (Night 3). The California Banana Slug, *Ariolimax columbianus*, returns to a depression it has excavated in the soil, and can do so after a foray to a point 4.5 metres away (Ingram and Adolph 1943). The Giant Garden Slug, *Limax maximus*, can return to its home site by a direct route from up to 90 cm away; slime-trail following is not involved (Gelperin 1974).

Students of ants and termites are more impressed by the trail-following behaviour of their animals and have been inclined to construct ethological models to explain all of the ant's foraging activities in terms of reflex

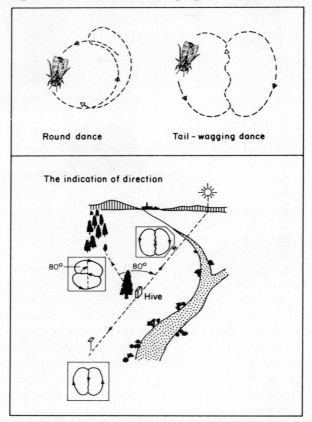

Fig. 6.6 The round and tail-wagging dances of the Honey Bee, *Apis mellifera*
The round dance is used to announce resources discovered near (less than 80 m, depending on subspecies) to the nest.
The waggle dance is used to announce resources discovered at greater distances. Direction is indicated by the direction of the run during the middle part of the figure-of-eight, the part during which the abdomen is waggled. On the vertical surface of a comb inside the hive the dancer transposes the angle of the resource relative to the Sun into an angle relative to gravity as in the bottom figure. The distance of the resource from the nest is indicated by the tempo of the tail-wagging dance.
[From Baker (1978a), after Frisch. Photo by Axel Doering, courtesy of Frank W. Lane]

responses to, and blind following of, chemical trails (see Schneirla 1971). In many ways, this inclination was a reaction to earlier authors who described the food-searching and slave-making forays of various species in the same terms as the strategic warfare of humans: platoons or regiments, their behaviour integrated by patrolling officers, performing pincer movements and sequential attacks to surprise and overpower their 'enemy'.

Army ants alternate phases of nomadism with phases of having a fixed base (Fig. 6.5). During the nomadic phase, a temporary bivouac is established each night under a log or similar site. A few raiding trails are established each day and along those food is carried back to the workers that remain in the bivouac to tend both the queen and the hungry, fast-growing larvae. At some time during the day, the colony decides where to migrate and establish a new bivouac during the coming night. When the developing larvae are large, voracious, and nearing pupation, the foraging workers lay down caches of food along the anticipated migration route so that the larvae can be fed *en route* during the coming night's migration (larvae are carried during migration in the mandibles of an ant; the queen migrates by her own locomotion). Each day, these ants must reach a decision concerning the site of the next bivouac, a site that is always located at a point on one of the raiding trails of the previous day. An ethological model would probably seek random variation in the amount of chemical trail deposited on various potential sites; from such imbalance, support would increase exponentially for the site on which most chemical was laid. A model in the spirit of behavioural ecology, based on optimal foraging, would seek an explanation other than statistical chance for one site being labelled more than others. The expectation would be that the ants on each raiding trail would collect information concerning potential bivouac sites and the relative density of prey in the surrounding areas. Ants from different trails would then compare information, perhaps visit each others' sites, and eventually reach a decision concerning the best site. The decision would then be communicated to at least some other members of the colony so that caches of food could be deposited.

Such a sequence of events seems beyond the ability of an animal that spends its life blindly following chemical trails. Many people would consider the whole idea of collecting information, making judgements, comparing those judgements with similar judgements made by other individuals, and then reaching a democratic decision, to be far-fetched, beyond the ability of even a non-human mammal let alone an insect. Were it not for the Honey Bee, *Apis mellifera,* and the extremely elegant work of Lindauer (1955), behavioural ecologists may have been forced to keep their contemplation of such heretical concepts to themselves. Lindauer has provided us, however, with incontrovertible evidence that insects are capable of such behaviour.

There can be few students of biology that are not familiar with the dance-language of the Honey Bee (Fig. 6.6) and the classic work of Karl von Frisch (1967). Less widely known, surprisingly in view of its significance, is the work of Lindauer on the mechanism by which a swarm of bees reaches a decision concerning where to migrate and settle in order to establish a new colony. On leaving the hive to migrate to a new site, a swarm consists of the old queen and about half the workers; the other half of the worker force and a new queen, which may not yet have emerged from the pupa, are left behind. Soon after leaving the parent site, the migrant swarm settles nearby as a hanging cluster. Within fifteen minutes of the cluster having formed, 'scout' bees start to dance on its surface.

I shall avoid controversy over what and what is not 'language' (Gould 1975, Rosin 1978 and Desmond 1979). The important fact is that by a combination of sound and movement one Honey Bee can communicate to another the distance, direction, and quality of a resource. In dancing on the surface of the cluster, the scouts are continuing a process that was begun up to three days before the swarm departed from the parent hive. This is the time that scout bees first become active and start to explore for potential 'nesting' sites. After discovering a potentially acceptable site, the scout bee returns to the hive and starts to dance, communicating information about the site to other individuals and perhaps recruiting them to join the swarm when eventually it leaves the hive. By the time that the hanging cluster has formed, a number of potential destinations have been discovered and the process of deciding to which site the swarm should migrate begins.

At first, different scouts announce different potential nesting sites. Each dancing scout recruits fresh bees to examine the site being announced. At intervals, the scout returns to this site and, using pheromone, attracts to the precise spot other scouts that are in the general area after having been

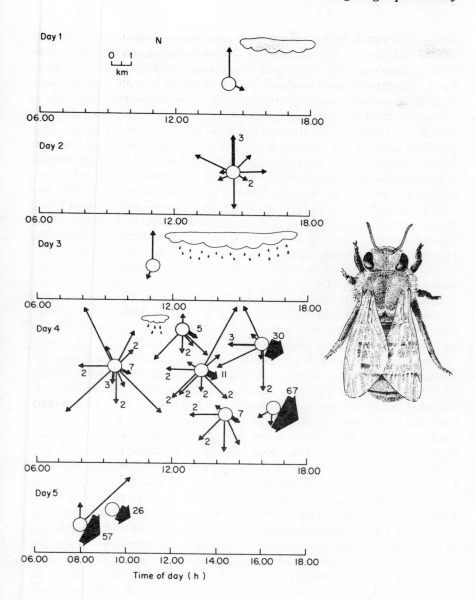

Fig. 6.7 Variation in support for different potential nesting sites during the period between emergence of a Honey Bee swarm from the hive and migration to a new nesting site The diagram shows the number of new recruits dancing on the surface of a Honey Bee cluster in support of different potential nesting sites. The length of each arrow shows the distance from the cluster of the potential nesting site being advertised. The width of the arrow and the number printed at the arrow head indicates the number of new recruits since the previous observation that are dancing in support of that site. Where no number is printed, only a single individual is involved. This particular swarm of bees required five days to reach a decision as to the most suitable destination.
[From Baker (1978a), after Lindauer]

recruited by her dancing. The number of new bees recruited to examine and in turn dance in support when they return to the cluster depends in part on the intensity of the dance performed by the origional scout and in part on the assessment of the site by the various potential new recruits during their visit. However, even once a bee starts to dance in support of a new site, it can change allegiance and support some other site after comparing the two. This happens when a bee, dancing in support of a moderately suitable site, perceives another bee dancing more intensely in support of another. On such occasions it will run after that bee and then go and inspect the site being announced. If the new site is judged to be more suitable than her original site, the bee, on returning to the cluster, dances in support of the new site. By this process, support for the different sites fluctuates while a final decision is being made (Fig. 6.7). Eventually, however, all scout bees dance in support of the same site. Not until this situation is reached does the cluster depart. Once the scouts have reached agreement, they perform a series of buzzing runs over the surface of the cluster. Within a few minutes, the whole mass is alerted and takes off. The swarm cloud proceeds gradually, but, within the swarm, a few hundreds bees, presumed to be the scouts, shoot ahead in the direction of the selected destination. Once they have cleared the swarm they drift slowly back before repeating the process. In this way, direction is conveyed to the entire swarm.

Once the decision-making process was known, the way was open for Lindauer to determine what cues the bees were using to assess the suitability of different sites. By watching the dancing bees, Lindauer could visit and find the sites being announced. He could then change their characteristics and observe whether support for the site, in terms of new recruits, increased or decreased. He found that temperature, smell, shelter, distance from the parent hive, and other factors were all taken into account.

We are faced, then, with the following facts. Honey Bees explore (see review of foraging by social insects by Heinrich 1978). During exploration, an individual visits a variety of sites and reaches a decision concerning not only which is the best but also how it relates to some absolute scale of suitability. According to this assessment, the bee dances with a certain intensity in support of 'her' site. She is receptive to greater intensity shown by another bee and is prepared to visit sites that are being claimed to be more suitable than her own. Upon visiting the site she is able to compare its suitability with the memorised suitability of her own site. This comparison is based, not on a single factor, but on many interacting factors. On the basis of this cerebral comparison, a decision is reached concerning which site to support on return to the cluster. Finally, the scouts are able to recognise when they are all in agreement about which is the best site. On top of this, there is the spatial memory necessary for exploration and navigation to occur. In what way are the cerebral processes involved in this behaviour fundamentally different from the processes by which a human decides where to live?

7

Exploration and communication

The Honey Bee provides a well-documented illustration of the way in which exploration can interact with social communication in the formation of an individual's familiar area. Were it not for such a spectacular example *in an invertebrate*, we might have been tempted to propose that the involvement of social communication was a factor which might set the human lifetime track apart from that of other animals. As this temptation has been forestalled, we can only examine situations in other species to see how widespread this involvement may be.

We know, from humans, that social communication can be involved with exploration in two ways, information being gained from another individual either through some form of language or by simple association or observation. During human childhood, independent exploration only takes place over relatively short distances. Most familiar-area establishment takes place in the company of other individuals, primarily parents or other adults. The familiar area thus established then serves as the base from which longer distance explorations are carried out in adolescence and later life.

Establishment of a basic familiar area during early life by associating with parents or other adults followed later by independent exploration is probably characteristic of nearly all mammals. Indeed, some group-living species seem never to perform solitary explorations. Examples are those female primates and ungulates that either never leave the female coalition (*sensu* Wrangham, personal communication, 1978) containing their mother or only join another group when the two groups come into contact. Other examples are those species, such as Wildebeest, *Connochaetes taurinus*, that live permanently in herds. If baleen whales, as suggested by Payne and Webb (1971), can really communicate by sound over distances of hundreds, or even thousands, of kilometres, then it seems likely that they also are rarely 'alone', despite the low level to which their populations have been decimated. The nearest any young mammal seems to come to building a familiar area by independent exploration without the benefit of a basic familiar area gained by association with adults is found in rodents, bats and pinnipeds. In these, the young of some species are weaned and become independent very quickly, within weeks or months of birth. Even so, young rodents may be 'shown' the mother's home range by their parent and many, such as Muskrats, *Ondatra zibethicus*, and Woodchuck, *Marmota monax*, may then establish a temporary roost near the edge of this range. This roost is used as a base from which to carry out their own independent explorations before deciding where to live. Young Prairie Dogs, *Cynomys*

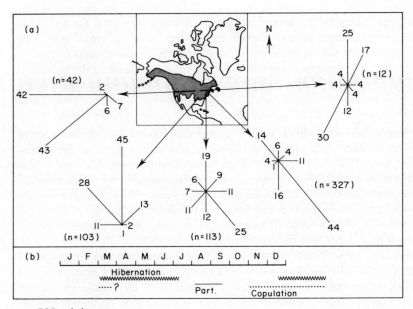

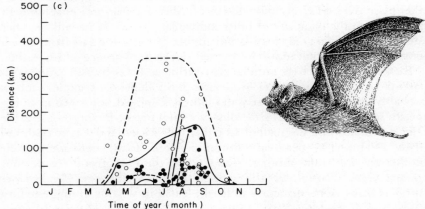

Fig. 7.1 Seasonal return migration of the Little Brown Bat, *Myotis lucifugus*
(a) Distribution and direction ratio. Stippled area, distribution. Direction ratios are shown for groups hibernating in Illinois, Kentucky, New York State, Vermont and Ontario as shown. Length of line and number shows the percentage of each group that was recaptured in summer in each of eight directions from the hibernation site in which the bats were marked.
(b) Seasonal behaviour. Solid line, parturition period.
(c) Distance from hibernation site with time of year. Solid dots, males; open dots, females. The maximum and minimum limits are indicated: solid lines, males; dashed lines, females. Straight lines connecting some recaptures in July, August and September relate to individuals marked during their summer visit to their hibernation site.
[From Baker (1978a), after Griffin, Humphrey, Cope, Davis, Hitchcock, Barbour, Fenton and Walley]

ludovicianus, may be presented with the parental territory for their first winter, the parents going elsewhere. Female insectivorous bats may accompany their young on the first stage of their post-weaning migration in order briefly to show them some of the roosts available in the area. This could be one (but not all—Baker 1978a) of the reasons for the often long-distance (200 km or more) post-breeding migrations made by some female bats when they visit their winter roost for just an hour or so before going back to their summer range (Fig. 7.1) and leaving the young thereafter to explore independently.

Most young birds associate with their parents for at least a short period after fledging, exceptions being parasitic species, such as the Cuckoo, *Cuculus canorus*, in which the eggs are laid in the nest of another species. During this association, the young build up a basic familiar area within which at least some (e.g. Reed Warbler, *Acrocephalus scirpaceus*) often find their future breeding site. It seems likely, however, that most young of most birds find their future breeding site during their independent post-fledging explorations. In some, such as geese and swans, the young stay with their parents as a family party during virtually their entire first year of life. Some Cranes and probably many other birds do the same. By the time they become independent, the young of such species have a basic familiar area that reflects the entire year's home range of their parents. The ability to recognise individuals is likely to be strong, though not necessarily greater than that of other species. Parents and offspring of Bewick's Swans, *Cygnus bewickii*, have been shown to be able to recognise each other even after they have (probably) been separated for six months or so.

Spectacled Caiman, *Caiman crocodilus*, stay together as a 'pod' for the first 18 months of life and are protected by parents and other adults (Gorzula 1978). They then begin a phase of independent exploration, during which they visit a wide range of lagoons. There has to be a possibility of parental involvement in the early phase of this movement. Most reptiles, amphibians, fish and invertebrates, however, are abandoned by their parents as eggs or young larvae and have no alternative but to begin independent exploration immediately upon hatching. Sundanese Rock Pythons, *Python bivittatus*, return from forays to roost in the egg shell from which they hatched, suggesting that exploration and the establishment of a familiar area begins immediately upon hatching. Larval amphibians rapidly learn the direction of the inshore/offshore axis of the pond in which they were spawned. Except for a few species (e.g. spiders that carry their young on their back for a few days), there can be no question of parental involvement in even the earliest stages of exploration by animals other than mammals or birds. Rather, if association with other, more experienced, individuals is involved at all in establishing a familiar area it seems to occur during the period immediately before first reproduction. Nearly mature Cod, *Gadus morhua*, for example, associate with an adult shoal for over a year before they themselves begin to spawn.

In general terms, therefore, young mammals and birds first establish a basic familiar area by association with adults and then use this as a starting point from which to perform their own independent explorations. All other animals that live within a familiar area begin immediately with independent exploration. If association with adults is involved at all, it occurs in the later stages of crystallisation of the adult home range.

So far, we have considered the involvement of social communication with exploration only in the sense of association with other, more experienced, individuals. As we know, humans also convey information concerning the distance, direction and characteristics of a location, and make comparisons with other locations, by word-of-mouth, writing, drawing, and movement. Individuals often seek as much of such information as they can from others before deciding where, or even whether, to carry out their own independent explorations. Exploration is still necessary, however, not least because such socially-communicated information is not always totally accurate. In any case, even if it is accurate, it is still no substitute for direct experience. Nevertheless, as by its very nature exploration is inefficient, even hazardous, such information is often invaluable.

Knowing that animals as different as Man, Chimpanzees (see Desmond 1979), and Honey Bees are able to convey information concerning places and routes by means of sound and movement, we should be alert to the possibility that other animals share such an ability. In the absence of critical data, there seems to be no harm in indulging in a little speculation, though many behaviourists will doubtless find the suggestions 'fanciful'.

Before hunting, many social, carnivorous mammals indulge in ceremonies involving vocalisation, gesture and movement. Wolves, *Canis lupus*, for example, circle round, rub noses, wag tails and vocalise before setting off to hunt. So, too, do African Hunting Dogs, *Lycaon pictus*. The temptation to conclude that a decision is being made about where to hunt and what strategy to use is strong. The loud vocalisations of Wolves and Coyotes, *Canis latrans*, (Lehner 1978) could also be a means of exchanging information about places and routes over long distances. A similar role could be ascribed to some of the noises made by baleen whales. Examples are the 20 Hz sounds made by Finback Whales, *Balaenoptera physalus*, (Payne and Webb 1971) and the well-known 'song' of the Humpback Whale, *Megaptera novaeangliae*. Herds of wildebeest, *Connochaetes taurinus*, frequently have to decide where next to migrate in their ceaseless search for fresh pasture on the plains of Africa (Fig. 7.2). The milling around that often occurs before departure could well be part of a decision-making process concerning when and where to go.

It is almost a decade since Ward and Zahavi (1973) suggested that the communal roosts of many birds might be centres for the exchange of information. Behavioural ecologists among ornithologists have been predisposed to accept the idea from the beginning but critical data are

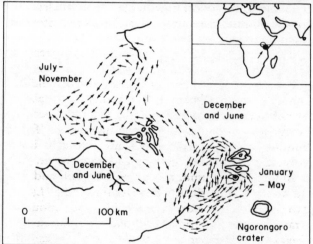

Fig. 7.2 Annual migration circuit of Wildebeest, *Connochaetes taurinus*, in the Serengeti region of Africa
Arrows show the predominant direction of movement. The herds have dry-season (July–November) and wet-season (January–May) ranges. Much movement, however, is entirely opportunistic and, particularly at the beginning of the wet season, is directed to any area where rain can be seen or heard to be falling at distances of up to 100 km.
[From Baker (1978a), after Grzimek and Grzimek and Talbot and Talbot.
Photo by Norman Myers (courtesy of Bruce Coleman Ltd)]

elusive and the suggestion is still not validated. Most authors envisage a more or less parasitic mechanism for information exchange. Birds that have fared less well on a given day observe the directions from which plump, well-fed, birds arrive and then either follow those birds or fly in those directions the next day. Observations at a large communal roost (up to

30 000 birds) of five gull species at Audenshaw Reservoir, Manchester, suggest a more positive mechanism. The thesis is that birds that have had a bad day announce the fact by a display flight as they arrive at the reservoir towards dusk. As they arrive, displaying birds fold their wings and dive down toward the water, often from a considerable height. As they do so, they frequently 'rock' their bodies from side to side several times, often vocalising at the same time. The model to be tested is that the display conveys the distance and direction of unfavourable conditions, distance perhaps being conveyed by number of 'rocks'. More individuals perform display flights on cold, frosty days in winter than at other times (J. Davies and Baker, unpublished), but so far critical data have not been collected.

Birds, such as gulls, starlings, and swallows, that roost or breed communally, often perform spectacular, communal, pre-roost display flights (Fig. 7.3). These could advertise the location of the roost to individuals, such as explorers, that would otherwise be unaware of the roost's existence.

A number of birds also collect together in late summer before setting off on a stage of the autumn migration. Barn Swallows, *Hirundo rustica*, for example, often aggregate on telephone wires. The behaviour shown is

Fig. 7.3 Starlings performing pre-roost display flight
(Photo by S. C. Porter, courtesy of Bruce Coleman Ltd)

instructive. Birds collect together, vocalising as they do so. Over the course of 5 minutes or so the group grows in size and the noise becomes louder. Suddenly they all take to the air, streaming away from the wires, before milling around, beginning to settle, and then repeating the entire process. Observations in Wiltshire, England, have shown that (1) the groups are a mixture of adults and young; (2) the initial direction in which they fly off, before milling around and resettling, is invariably south-east, the standard autumn direction for the species in the region. More data over a wider area are needed; but could it be that adults are teaching the young the appropriate direction for migration (see also Chapter 14), or perhaps teaching them to compensate for the Sun's movement across the sky?

Prior to 1970 behaviourists were reluctant to accept the existence of any behaviour that seemed to confer even a short-term disadvantage on the individual concerned. The only theoretical model purporting to explain the evolution of such behaviour was that of group selection, formalised in the justly famous book by Wynne-Edwards (1962). Many people disliked group selectionism but had no model with which it could be replaced to explain the evolution of altruism. Now that the group selection heresy has been superseded by the models of reciprocal altruism (Trivers 1971) and Kin selection (Hamilton 1963, Maynard Smith 1964, Dawkins 1976) (see brief reviews by Clutton-Brock and Harvey 1978, and Parker 1978) acceptance of apparently altruistic behaviour is no longer a problem. Indeed, the only consideration is usually which model is most appropriate to any given situation and behaviour.

I am inclined to support the view of Dawkins and Krebs (1978) that all social communication is manipulation rather than simple transfer of information. Sometimes the information conveyed by an individual is accurate, sometimes it is exaggerated, and sometimes it is false. Almost always the information imparted is that considered by the individual providing the information to be in its own long-term interest (given the framework of reciprocal altruism and/or kin selection). This will be so whether or not the animal concerned is human. In particular, it will be true in the case of information conveyed to a would-be explorer. Depending on whether it is to the resident's advantage for the explorer to go or stay, settle here or there, go this way or that, will depend the information given and its accuracy. Equally, the explorer may or may not be inclined to believe the information given.

The picture emerges that, in travelling, an explorer continually has to process information that is offered by, or sought from, residents, other explorers, and other individuals in general. In effect, exploration takes place against a backdrop of the distribution of all other individuals. This backdrop has an important influence on the pattern of exploration and is the subject of the next chapter.

8

The territorial backdrop

We have concluded that all vertebrates and many invertebrates organise their lifetime track in such a way as to live within a familiar area and make use of a sense of location that is just as cerebral as that of humans. This sense of location is achieved by a process of exploration during which judgements are made, and acted upon, about places to live and exploit. In the last chapter we saw the way that some would-be explorers seek information from other individuals. As an explorer travels, he is forever coming across conspecifics which offer information, exaggerating and trying to manipulate his behaviour. Some may invite the explorer to stay but most will urge him to leave; to settle elsewhere. In effect, the distribution of conspecifics forms a territorial backdrop against which each individual's lifetime track is played out. In order to appreciate fully the process of exploration, we have to give some thought to this territorial backdrop and the influence it has on the lifetime track.

In the discussion that follows, I shall adopt the use of the term 'territory' suggested by Davies (1978): territories exist whenever individual animals or groups are spaced out more than would be expected from a random occupation of suitable habitats. Animals that establish a familiar area are affected by the territorial backdrop in three major ways: (1) access to resources during exploration; (2) where and how the explorer eventually attempts to settle; and (3) the size and form of the home range eventually established.

As far as access to resources during exploration is concerned, a migrant has the choice of exploiting, or attempting to exploit, either occupied or unoccupied areas. Unoccupied areas may be deficient in particular resources but often one resource (e.g. food) may be available in an area that has no permanent resident because it is deficient in some other resource (e.g. shelter). An explorer, travelling through a larger area than would be optimum for a settled individual, may be able to maintain itself entirely on resources obtained in areas unsuitable for permanent occupation. Some resources, however, may have to be obtained from occupied areas. Food, for example, could be obtained from an occupied home range if the explorer could avoid detection by the resident. This might be another reason for inconspicuousness during major phases of exploration, perhaps as important as the avoidance of detection by predators (Baker and Parker 1979). Alternatively, during periods of resource abundance, it may not be to the advantage of a resident animal to spend time and energy defending a resource against explorers. Thus, Black, Kodiak and Grizzly Bears aggregate at salmon streams (Fig. 8.1), Orang Utans converge on heavily fruiting trees, and Rats feed in large numbers on rubbish tips.

Given that an explorer is able to obtain sufficient resources to maintain itself, it eventually has to solve the problem of finding the best place to attempt to establish a home range of its own. The best strategy to adopt is likely to depend on deme density (Baker 1978a, Fig. 18.25). If deme density is low and unoccupied areas are suitable for permanent occupation, the explorer should settle in the best of the unoccupied habitats. If deme density is moderate and unoccupied areas are unsuitable for permanent occupation, the explorer should attempt to infiltrate an occupied area and carve out a home range for itself between the home ranges of established individuals. Finally, if all areas are occupied at maximum density and there is no room for infiltration, the explorer should either attempt to oust an established individual and take over its home range or subsist in the best of the sub-optimal habitats and try to be the first to be on hand to take over an area when an established individual becomes ill, dies or leaves.

Consider two adjacent areas, A and B. Area A is supporting 5 residents and B is supporting only 1. In which area should an explorer eventually attempt to settle? Clearly it depends on the resources also available in each area. If a resources are available in A and b resources in B, the critical question is whether $a/5 + 1$ is greater or less than $b/1 + 1$. Whichever is greater, the migrant should attempt to settle in that area. Suppose the first migrant infiltrates into A. The critical judgement for the next migrant is whether $a/6 + 1$ is greater or less than $b/1 + 1$. It can be seen that in such a situation, termed a *free* situation by Fretwell (1972) because explorers are free to settle where they choose, animals eventually settle in the two areas at such densities that the resources available to each individual in each area are equal. Hence, in a free situation all individuals gain equal benefit from their behaviour.

As deme density per unit resource increases, the situation becomes one of increasing despotism. A *despotic* situation (Fretwell 1972) is one in which some individuals, by virtue of a high *RHP* (*resource holding power*; Parker 1974b), exclude from resources other individuals of lower RHP. RHP is influenced, not only by innate factors such as size, strength and cunning, but also by age, experience and location, as well as by chance factors such as the number of disease organisms contracted by the individual and the amount of physical damage so far sustained.

Individuals of different RHP exist in a free situation as well as in a despotic situation. In a free situation, however, it seems likely that because of the lower deme density, the benefits that individuals of high RHP would gain by excluding low RHP individuals would not be worth the effort—the time, energy, and risk of being wounded. As deme density increases, however, the benefit gained from the physical defence of resources increases and there eventually comes a point at which it is worth the cost (effort and danger). At this point, high RHP individuals should begin

Fig. 8.1 Kodiak bear, *Ursus arctos*, catching salmon. (Photo by Fred Bruemmer)

physically to prevent low RHP individuals from gaining access to resources. In a free situation, high and low RHP individuals do equally well. With increasing despotism, the difference in reproductive success between high and low RHP individuals increases and in extreme situations low RHP individuals fail to reproduce at all.

When deme density is very low, there is an advantage to the explorer in being able to detect which areas are already occupied and which are not.

Fig. 8.2 The Great Tit, *Parus major*, has a number of songs and sub-songs in its repertoire and often switches from one to another at the same time as it changes position to a new branch or tree. (Photo by Eric Hosking)

Such knowledge makes it easier to judge where it is best to settle. At the same time it is an advantage to each resident to advertise its presence in an area so that incoming explorers are less likely to attempt to settle within the resident's home range. When deme density is moderate and the situation is free, it is an advantage to the explorer to be able to detect the number of residents in each area, again in order to judge where best to settle. It is still an advantage to the residents to advertise their presence. Now there would seem also to be an advantage in 'cheating' by exaggerating, if possible (Fig. 8.2), the number of individuals that seem to be present in the resident's home range; the so-called 'Beau Geste strategy' (Krebs 1977) after the general who propped up his dead soldiers so as to give the impression of greater numbers. When the situation is despotic, there is an advantage to the explorer in being able to detect which territories are currently occupied, where the boundaries lie, and, if possible, the relative RHP's of the occupants. There is also an advantage to the territory holder in advertising its continued presence and perhaps also in exaggerating its RHP (Parker 1974b).

It follows that no matter at what position a deme may be on the free—despotic territorial spectrum, there is always an advantage to both residents and explorers in the residents advertising themselves. It seems also to be an advantage for the residents at times to be selective in the information conveyed by their advertisements and often to exaggerate—in other words, to attempt to manipulate the behaviour of the explorer in a way beneficial to the resident (see Dawkins and Krebs 1978). At the same time, explorers should try to see through any such attempted deception and to interpret the advertisements as accurately as possible. The explorer is likely to seek information from the resident about the locations being visited and exploited, time since last visit, gender (of the resident), physiological state, RHP, and individual identity. It is in the resident's interest to give such information, albeit selectively and with appropriate exaggeration. We should not expect, therefore, that the territorial backdrop will be difficult for explorers to detect. On the contrary, within the confines of predation pressure, we should expect selection to have favoured resident animals conspicuously advertising all of these features and explorers to be finely tuned to detect them. To an explorer, the territorial backdrop is probably the most vivid feature of its environment.

The response of the incoming explorer upon receiving such information will depend on a number of factors. If deme density is very low, an explorer is likely to leave rapidly any area that is already being used. In a free situation, it will not necessarily leave the area but the information will be used to judge deme density and the best place to attempt to infiltrate and settle. In a despotic situation the information will be used to assess RHP before making, or avoiding, physical contact with the territory holder. The explorer may be able to detect from the residents' advertisements which individuals are most likely to decrease in fitness. Finally, the observed failure of a territory holder to advertise its presence after the usual interval could give an explorer its first indication that a territory has fallen vacant.

In view of the subtle range of information-transfer, manipulation and response associated with scent marking, vocalisations, and other forms of advertisement, to brand them simply as 'territorial' is a gross over-simplification. At low deme densities, because incoming individuals avoid occupied areas, advertisement by residents may be observed to deflect migrants from the area. In a free situation, precisely the same form of advertisement will deflect explorers from high-density areas but will actually attract them to lower-density areas. In a despotic situation the same form of advertisement will again deflect explorers, unless a migrant judges the advertiser to be of lower RHP than itself in which case it will be encouraged by the display to attempt to stay. Whatever the situation, the explorer is likely to travel around and collect information before deciding where to try to settle.

The influence of the territorial backdrop does not end with the decision by the explorer concerning where to attempt to settle. It also affects the form of the home range that is eventually established.

When deme density is very low and the explorer settles in the best of the unoccupied habitats, there should be no social constraints on the size and form of the home range eventually established (except insofar as males should preferentially select a home range that gives them access to females and *vice versa*). Size of home range (see review by Davies 1978) should be determined by the best 'trade-off' between the increased cost of moving within and maintaining familiarity with a larger area and the resulting

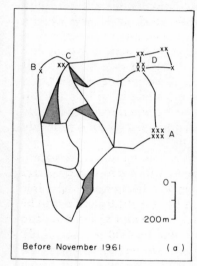

Before November 1961   (a)

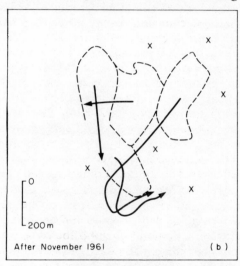

After November 1961   (b)

Fig. 8.3 The effect of deme density on the territorial system of the Weasel, *Mustela nivalis*

(a) Territories of male Weasels in a young forest plantation in the Carron Valley, Stirlingshire, during the winter (November through March) non-breeding season. Heavy lines outline the territories. Hatching shows areas of overlap or where edge of the territory could not be determined. Crosses show the history of a male yearling Weasel that was caught only at A from November 1960 to January 1961 and then within a few days appeared at B, C and then D where it took over a territory that had just become vacant.

(b) Movements of surviving territory occupants following a sudden decrease in deme density. Crosses show territories that fell vacant in November 1961. The territories (dashed lines) of the surviving males did not expand but instead gradually shifted in the directions shown by the heavy arrows. Shifts were largely in the direction of neighbouring, unoccupied areas. This mobile system persisted throughout 1962 and 1963.

[From Baker (1978a), after Lockie]

increase in available resources. Any contact or overlap with the home range of a neighbouring individual of the same sex is a disadvantage if resources in the zone of contact are shared. The result should be a withdrawal of the home range from the zone of contact and perhaps a compensatory expansion in the opposite direction. The mobile situation found in one study of the Weasel, *Mustela nivalis*, (Fig. 8.3) and observed gradual shifts in home range by other species (Fig. 3.5) could well reflect this type of situation.

In a free situation with moderate density the incoming migrants invariably settle within a fully occupied area and establish a home range at the expense of neighbouring individuals. The resultant home range, therefore, has a zone of contact or overlap with those of neighbouring individuals all round its perimeter. At such a low density it is not worth spending time and energy preventing neighbours from gaining access to

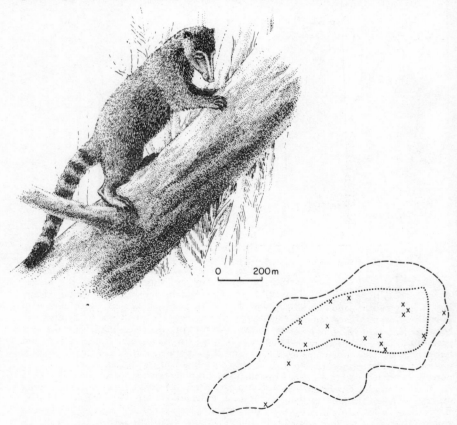

Fig. 8.4 Home range (dashed line), core area (dotted line), and roosting sites (x) of a social group of the Coati, *Nasua narica*, on Barro Colorado Island, Panama. The core area is used for about 80 per cent of the time.
[From Baker (1978a), after Kaufmann]

peripheral resources. As a result, because they are shared, these will be reduced in quantity and be less constant and predictable in their availability than more central resources. Peripheral habitats should therefore be ranked as less suitable and consequently calculated and RFR migrations to such sites will be less frequent than to more central sites. The final result is that the animal spends far more time in the central part (or *core area*) of its home range than in the areas of overlap with neighbouring home ranges (Fig. 8.4).

As deme density and despotism increase, other individuals can only be prevented from gaining access to resources in the central area of the home

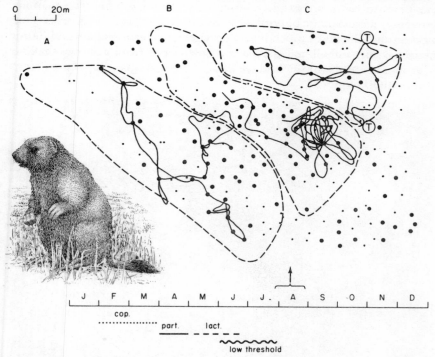

Fig. 8.5 Territories and track pattern of the Black-Tailed Prairie Dog, *Cynomys ludovicianus*, in the Black Hills, South Dakota

Seasonal behaviour is shown at the bottom of the diagram. cop., copulation; part., parturition; lact., lactation; waved line, period that adults are most likely to leave their territory to establish a new one elsewhere. All territories shown were occupied by social groups; track patterns recorded in August. The large and small dots indicate, respectively, burrows with and without a crater and mound.

A. Track of an adult female during a 286 min period. The territory was occupied by 1 adult male, 4 adult females, and 14 young of the year.

B. Track of an immature male during a 430 min period. The territory was occupied by 31 young of the year, without adults.

C. Track of an adult male during a 180 min period. The territory was occupied by this male, 2 adult females, and 8 young of the year. The encircled letter T shows where territorial interactions took place.

[From Baker (1978a), after King]

range by physical aggression or the threat of such aggression. At intermediate density levels there may still be zones of overlap outside of an aggressively defended core area. Beyond a certain density, however, the territorial individual can spend little time outside of its territory. Zones of overlap, and hence a core area, disappear (Fig. 8.5). Indeed, as the animal has to spend so much time defending and patrolling the boundaries of its territory, more time may be spent around the periphery of the home range than centrally.

In a despotic situation, at least during periods of intensive territorial defence, settled individuals have little time or opportunity for further exploration, using the existing home range as a base. In free situations, however, there may be opportunity for exploratory and RFR migrations to continue, albeit at a much reduced frequency, even after an adult home range has been established. This is particularly possible for species that live in groups because some individuals can continue to occupy the home range while others explore.

Fig. 8.6 A territorial alliance in action: New Guinea Highlanders. (Photo courtesy of Film Study Center, Harvard University).

Most discussions of territorial behaviour emphasise the individual nature of the territories. I have implied here that neighbouring residents could in addition often benefit from functioning as a unit, manipulating explorers with information in an attempt to encourage them to settle elsewhere. Once an individual has established a home range and become a part of the territorial backdrop it will often have little to gain from further competition with neighbours but much to gain from the manipulation of explorers. In such a situation, there would seem to be an advantage on the basis of reciprocal altruism, or occasionally kin selection, in entering an alliance with neighbouring territory holders jointly to manipulate explorers to go elsewhere. Although the Beau Geste principle was formulated for individuals (Krebs 1977), we can also apply it to all the residents in an area as a unit. If, by their joint action, they can give the maximum impression of a high deme density, explorers could well be encouraged to search elsewhere without investigating individual home ranges. Perhaps this is the explanation for the fact that neighbouring territory holders among birds often synchronise their advertisement displays. Carrying this theme a stage further, there could also be an advantage for territory holders (again on a reciprocal basis), to announce the arrival and passage of explorers (and perhaps predators) for the information of neighbours. In part, this may be the explanation for the otherwise redundant complexity of many bird songs.

Perhaps, then, the territorial backdrop is not simply made up of individual home ranges. Instead it could consist of much larger unit areas, each one containing its own *territorial alliance* of established individuals. Although at times these continue to compete amongst themselves in border disputes in the classical sense, at other times each member of the alliance may benefit from cooperation to manipulate their common enemy, the intruding explorers (and predators).

So far, we have confined our discussions of exploration to short distance movements. We now have some appreciation of the way that animals with a more or less fixed home range when adult gain their sense of location and decide where to settle. The last and major question is whether this same behaviour can possibly explain the much more spectacular long-distance, seasonal movements of birds, fish and the other classical migrants. This is the subject area which felt the main force of the ethological models of the lifetime track; this is where the final judgement will be made of the universality of the anthropomorphic view of vertebrate migrations. This is the subject of the next chapter.

9

Exploration and long-distance migration: the final challenge

Consider Man. Presumably pastoral nomads have no less cerebral a sense of location than other, more sedentary, humans. Presumably, also, the processes by which a pastoral nomad or hunter−gatherer builds up his familiar area are no different from those of an industrialist. Only the form of the adult home range is different. In previous chapters we have concluded that all vertebrates and many invertebrates share Man's cerebral sense of location and establish their familiar area by exploration in an essentially similar way. Why, then, should we expect long-distance migrants of other animals to have any less cerebral a sense of location or to establish their familiar area by any means other than exploration? Yet that is precisely what the old ethological models did expect. Does the exploration model fit the known facts better than the classic clock-and-compass or goal-area navigation models for birds or the olfactory imprinting model for fish?

The main feature shared by the clock-and-compass and goal-area navigation models of bird migration (Fig. 2.1) is that the migration routes of adults are the direct result of genetic programmes already possessed when juvenile. Any change in migration behaviour by adults over a period of time, therefore, must reflect a change in the genetic programme and must also be shown by juveniles. Suppose, instead, that the autumn and spring migrations of young birds are exploratory, continuations of post-fledging exploration but with a preferred compass direction. As usual, the habitats visited are assessed and ranked. Suppose further that the environment changes so that in one generation the best winter habitats, say, are in the southern part of the range explored whereas in the next the best habitats are in the north. The exploration model would allow, even expect, the migration of adults to change accordingly. The migration patterns of young birds, however, should not change, or at least not as much. The total area still has to be explored before the bird can assess which areas are best.

If we could find a bird species that has recently changed its migration pattern, we could test whether the exploration model was really more acceptable than the purely ethological models. If it were found that adults had changed their migration pattern with no corresponding change in the migration pattern of first-year birds, we could accept the exploration model in preference to the other two. On the other hand, if the migration patterns of young and adults show similar trends, we should have to reject the exploration model in favour of a model based on programmed juveniles. The recent history of the Lesser Black-Backed Gull, *Larus fuscus*, in Western Europe provides us with just such an opportunity.

Over the past few decades, there has been a dramatic increase in the number of birds of this species that overwinter in Britain. Analysis of the ringing recoveries of *L. fuscus* born on Walney Island, England, shows that birds migrate overland across Britain and then travel along the coasts of France, Iberia and North-West Africa (Fig. 9.1). Overwintering adults in the years 1962–68 were found primarily from southern France

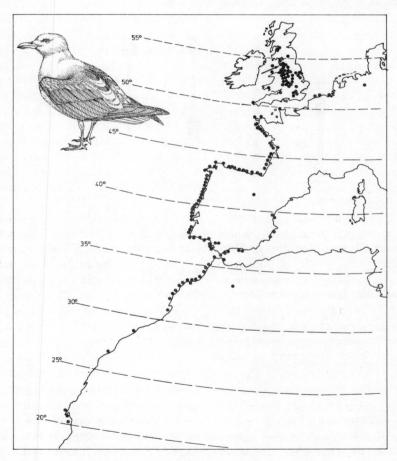

Fig. 9.1 Migration route of Lesser Black-Backed Gulls, *Larus fuscus*, born on Walney Island (arrowed) as indicated by ringing recoveries. Each dot shows the recovery of a dead bird ringed as a pullus on Walney Island between 1962 and 1975. The migration route is overland across England but seems to be coastal thereafter.
[Re-drawn from Baker (1980a)]

southwards. In the years 1969–75, however, they were found primarily in Britain (Fig. 9.2). During the period 1962–75 the rate of northward shift in winter range averaged more than one degree of latitude per year (Baker 1980a). The advantage to adults of wintering further north in recent years is unknown but may relate to the enormous population increase of this species in this century and a possibly increased advantage in being first to return to the breeding colony (Haartman 1968). Whatever the reason, the important fact in the present discussion is that the distribution of birds during their first and second winters has not changed over the years (Fig. 9.2). We may therefore accept the exploration model in this case in preference to the classical models.

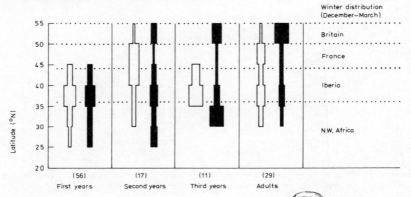

Fig. 9.2 Change in winter distribution of Lesser Black-Backed Gulls, *Larus fuscus*, born on Walney Island, England, as indicated by ringing recoveries
Histograms show the relative distribution of ringing recoveries, at intervals of 5° of latitude, for birds during each of their first three years of life and for adults. Figures in brackets show on how many individuals each pair of distributions is based. Open histograms, recaptures 1962–68; solid black, recaptures 1969–75. The distribution of adults has changed during the period whereas that of first- and second-year birds has not.
[Re-drawn from Baker (1980a)]

The *L. fuscus* analysis gives us even more information concerning the migration mechanism. Figure 9.3 compares the distribution of adult and first-year birds for each month of the year during the period 1969–75. The distributions are very similar from the end of breeding in July until November. Thereafter, the two distributions dissociate. It looks as though adults go with the young on a southward migration, only to return north in late autumn and early winter, leaving the young to begin independent exploration in the southernmost part of the range thus visited. There are two further pieces of evidence which support this view: (1) Adults migrate south more slowly now than in previous decades. So, too, do the young,

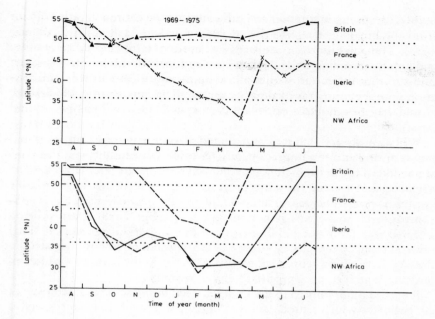

Fig. 9.3 Distribution of adult and first-year Lesser Black-Backed Gulls, *Larus fuscus*, for each month of the year.
All birds were ringed as pulli on Walney Island, England, and recovered dead during the period 1969–75. The top diagram shows the mean latitudinal distribution for adults (triangles) and first-year birds (crosses)
The bottom diagram shows the extreme limits for recoveries of adults (solid lines) and first-year birds (dashed lines). The distribution of adults and young is similar until November and then dissociates.
[Re-drawn from Baker (1980a)]

even though by winter the young have a distribution indistinguishable from their previous one (Fig. 9.2). (2) Harris (1970) carried out some cross-fostering experiments between *L.fuscus* and the closely related Herring Gull, *L.argentatus*, on Skomer Island, Pembrokeshire. Herring Gulls that breed on Skomer spend the winter along the coasts of the Irish Sea, not travelling south like Lesser Black-Backs. Young Herring Gulls and Lesser Black-Backs raised by foster parents, however, showed a migration pattern more similar to their foster species than their natural species. Evidently, association with parents (foster or natural) or observation of adults of the learned 'parental' species was an important part of the migration of these young gulls (when adult they were also predisposed to pair with a member of their foster species). At 5 months or so of age, however, young Lesser Black-Backs seem to begin a phase of independent exploration.

In geese and swans independent exploration may not begin until 9 months or more of age and after accompanying the parents on a more-or-less complete year's migration circuit. For Canada Geese, *Branta canadensis*, in North America, the first independent migration of the year-old bird involves retracing in summer and early autumn the entire migration circuit from the breeding range to the winter range and back again (Raveling 1976). This migration is completed just in time to turn round once more and join the breeding birds on their conventional autumn migration. Such mid-summer migration would allow opportunity for further familiaris-ation with the route travelled previously with parents, during which there would have been little opportunity for independent exploration and assessment.

When ducks are captured and ringed at transient roosts on lakes and ponds during autumn migration, any recaptures during the next few days are as likely to be back along, or sideways from, the southerly *standard migration direction* for the time of year. Such movements have become known as *reversal migrations* and are difficult to reconcile with clock-and-compass and goal-area navigation models (Rabøl 1978), usually being ascribed to a temporary malfunction of the migration mechanism. That reversal is only a temporary change of migration direction is suggested by the fact that ringing returns of, for example, ducks show that by winter all birds have arrived at appropriate winter quarters.

Reversal is not limited to ducks; it has been found to be characteristic of all species for which critical data are available and occurs in spring as well as in autumn, being detectable both visually and on radar (Alerstam 1978, Richardson 1978). In other words, it seems to be a consistent feature of the migration mechanism and must be accommodated by any realistic model of bird migration. Ringing returns suggest that reversal is primarily a characteristic of young birds and as such fits in beautifully with the concept of exploration.

According to the exploration model (Fig. 9.4), a young bird in autumn alternates migration in the direction standard for the species with sideways and reversal migrations. While migrating in the standard direction for the time of year, the young bird perceives distant habitats and ranks them provisionally according to suitability as transient home ranges. Reversal migrations are movements back along the track to visit and explore in more detail the most promising of those sites perceived earlier while flying overhead. Sideways migrations are in part similar, to visit habitats per-ceived to the side of the original track, and in part movements to locate and explore further sites over a wider area than is possible during the single flight in the standard direction. Eventually, as a result of these reversal and sideways migrations the bird should have located and ranked many of the suitable habitats in the area. These can then be used as transient home ranges for resting and feeding-up in future years.

Using as a springboard the familiar area established in the breeding range during post-fledging migration, the autumn migration of a young bird can be envisaged as a series of exploratory migrations, those in the standard autumn direction alternating with others back along and sideways from the standard track. The result is a gradual extension of the familiar area in a

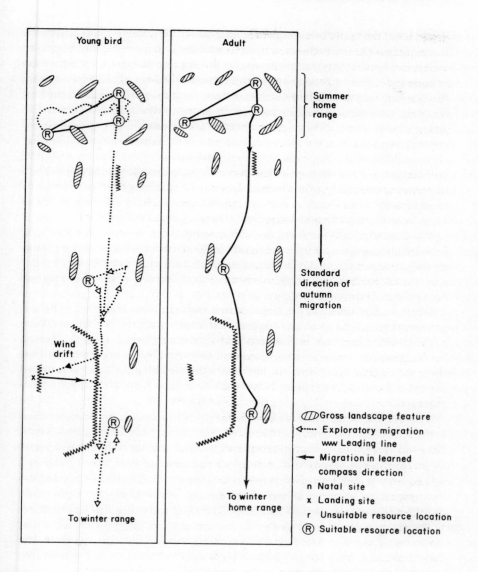

Fig. 9.4 The exploration model of long-distance seasonal migration by birds

During post-fledging migration the young bird explores widely in the breeding range, finding several resource locations such as potential breeding sites. During autumn migration, the familiar area is pushed in the direction standard for the species. Each migration leg in the standard direction alternates with movements back along and sideways from the original track. During these explorations, suitable transient home ranges are located for use as stopover sites in later years. The process is repeated until the bird arrives in its winter home range (Chapter 14). If wind drift occurs, it is countered by route-based navigation (Chapter 10). In future years the now adult bird migrates within the familiar area built up during its first year's explorations.

[Simplified from Baker (1978a)]

broad band from the breeding range in a direction standard for the species. It ends when the bird arrives in winter quarters (Chapter 14) and begins to search for suitable winter home ranges. Spring migration is a continuation of the exploratory process, often, though not always, taking place within the familiar area established during autumn migration. Either way, the best transient feeding and resting home ranges are unlikely to be the same in autumn as in spring. As a result, there is an advantage in continuing the exploration process until the bird arrives back (Chapter 10) at the familiar area established during post-fledging exploration the previous summer. Although this brief description of the exploration model has been couched in terms appropriate to temperate land-birds that migrate on a conventional north—south axis, it is easily modified to include the migrations of birds at sea and in tropical regions (Baker 1978a, Chapter 27).

Recent visual and radar studies of variation in the volume and direction of autumn and spring migration have shown that whether a bird migrates or not on a particular day or night depends to a high degree on weather conditions (Richardson 1978, Alerstam 1978). Even if a bird does migrate on a particular day, the direction of migration is also dependent on weather conditions. My own, as yet unpublished, visual observations of gulls migrating through Manchester suggest that adults wait for conditions to be suitable for migration in the standard direction (Baker 1978b). Young birds, however, migrate in the direction favoured by whatever conditions there are on that day. Other studies do not distinguish adults and young and so are difficult to interpret. Nevertheless, a few clear conclusions have emerged.

The most important factor to influence migration volume and direction is the speed and direction of the wind. Studies in Scandinavia and North America have shown that most species of birds are more likely to initiate migration in the standard direction for the time of year when there is a following wind (Richardson 1978, Alerstam 1978). They have also shown that migration in non-standard directions are more likely to occur when there is a following wind (Richardson 1978). Combining this information with my own observations of gulls, we can perhaps suggest that, whereas adults wait for the wind direction to be favourable for travel in the direction they wish to go, young birds take advantage of the wind to explore areas that happen to lie in the downwind direction.

This possibility leads us on to another line of thought. It has long been known that vagrants, birds that appear beyond the normal range of distribution of their species or race, etc., are invariably young, usually in their first year of life. Moreover, they usually appear at times that can be related to winds from their normal range to the area in question (K. Baker 1978). Previously, such appearances have been interpreted in terms of the birds being subjected to downwind drift (Williamson 1955, Lack 1960) and becoming lost. Another possibility now presents itself: that such birds are long-distance explorers, taking advantage of winds in a particular direction to explore areas that would not otherwise have been accessible. How can we test this notion? If it is true, vagrant birds should show some indication of not being lost; of having retained their sense of location. We can look for two features in their behaviour: (1) some indication that they return from whence they came; and (2) that they leave at a time that takes into account how far they have to travel. There are data that support both of these expectations. Unfortunately the eventual fate of vagrants has rarely been established. Few are ringed and the probability of recapture is low. However, an American Ring-Necked Duck, *Aythya collaris*, banded at Slimbridge, England in March 1976 was recovered in south-east Greenland in May 1977. A Rustic Bunting, *Emberiza rustica*, banded at Fair Isle in 1963 was recovered later the same year in Greece. Both birds showed evidence, therefore, of having returned, or being in the process of returning to, their normal range. Moreover, an analysis (Baker and Burns MS) of data presented by Sharrock and Sharrock (1976) shows a negative correlation for all categories of vagrants in Britain between mean date of occurrence in spring (January to May) and the distance to the nearest point of the breeding range for the species. Such a correlation suggests that the birds have, and are acting upon, information concerning their distance from the breeding grounds.

If vagrants are not lost but are long-distance explorers, why is it that they rarely settle in the distant areas that they visit? The implication is that they rarely consider areas distant from their normal range as being suitable. Perhaps this is mainly because of the absence of adults of their own species in such areas. If the nearby presence of adults is important for a habitat to be considered suitable, it would explain why, when a species expands its range, such expansion occurs gradually (Fig. 9.5) with few long-distance colonists, even though long-distance explorers may visit an area for many years before some eventually decide to breed.

The advantage of exploration is that it allows the individual to locate and exploit the best places that are available. As far as long distance migrants among birds are concerned, exploration during post-fledging migration identifies potential breeding sites. Similarly, exploration during winter identifies suitable winter sites, and exploration during autumn and spring migrations identifies suitable stopover sites for seeking shelter and food during future migrations. Perhaps further exploration for breeding sites also occurs in spring. It should certainly do so in individuals that do not breed until they are older than one year. Indeed, such species have the opportunity while still non-reproductive to explore prospective breeding sites while adults are actually in the process of breeding; the young can use the success of adults in the various sites as a measure of how good that site is for breeding. There is no shortage of evidence that the young of such birds (e.g. sea-birds, a few passerines) visit and 'prospect' a number of colonies and breeding areas while the adults are breeding.

If this view of autumn and spring migration is correct, we should expect

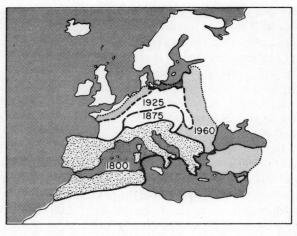

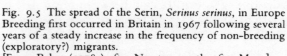

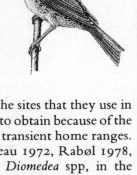

Fig. 9.5 The spread of the Serin, *Serinus serinus*, in Europe
Breeding first occurred in Britain in 1967 following several
years of a steady increase in the frequency of non-breeding
(exploratory?) migrants.
[From Baker (1978a) after Newton, partly after Mayr]

the best of the sites explored by young birds to be the sites that they use in
future years. Evidence for this has also been difficult to obtain because of the
short period that birds spend in autumn and spring transient home ranges.
However, data are accumulating rapidly (see Moreau 1972, Rabøl 1978,
Baker 1978a) and birds as diverse as Albatross, *Diomedea* spp, in the
Southern Ocean and Whinchats, *Saxicola rubetra*, in a Tunisian oasis are
being shown to revisit in later years transient home ranges used previously.
Most such evidence comes from trapping and ringing projects carried out
in communal roosts used by migrating swallows, wagtails and buntings,
etc. The most structured evidence, however, comes from the classic
experiment by Perdeck (1958) on Starlings, *Sturnus vulgaris*, that paradoxi-
cally was for many years the backbone of the clock-and-compass model
(Fig. 9.6).

Perdeck captured young Starlings during their migration through The
Netherlands to western wintering grounds and displaced them to
Switzerland. Recaptures the subsequent winter were all to the west and
southwest of their release point, many within the Iberian Peninsula. There
is still debate over whether the following spring these birds found their way
back to their normal breeding range or whether they arrived at destinations
east or east and south-east of this range. In my view, Fig. 9.6 suggests that
the birds did find their way back to their normal ranges. Either way, the
important point for this discussion is that, as would be expected according
to the exploration model, most birds in the following winter returned to
their Iberian winter home ranges rather than to the traditional winter range
in Britain and France. They did this despite the fact that they must have

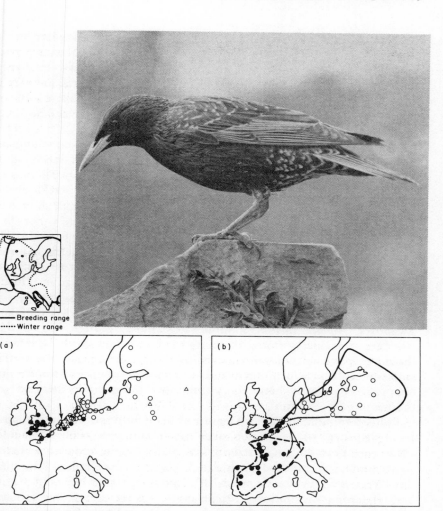

Breeding range
Winter range

(a)

(b)

O Breeding season (April to August)
△ Migration (March and September to November)
● Winter (December to February)

——Breeding range of controls
········Winter range of controls
— —Range in 1st winter after displacement

Fig. 9.6 The behaviour of young Starlings, *Sturnus vulgaris*, displaced experimentally from the Netherlands to Switzerland

(a) *Recapture of controls* Starlings were captured and ringed in the Netherlands (arrow) in October and November. Young birds (approximately 6 months old) were divided into two groups. Controls were released at the point of capture. Subsequent recaptures are indicated according to the season of recapture.

(b) *Recapture of displaced birds* The second group of young birds captured and ringed at the same time and place as the controls were transported by aeroplane to Switzerland. Continuous and dotted lines enclose the breeding and winter range of controls taken from (a). Recaptures of displaced birds during their first winter are not shown but the dashed line encompasses the total area in which recaptures were made. The recaptures shown, using the same convention as in (a), refer to birds recaptured in seasons and years after their first winter.

[From Baker (1978a), after Perdeck. Photo by Eric Hosking]

become involved on occasion in flocks of conspecifics travelling in the 'normal' direction. Those few that did arrive in the traditional winter range probably did so as a result of association with such flocks. Evidently, therefore, having explored the route from the Baltic to Iberia, individuals of even a bird as social as the Starling were inclined to make use of this familiarity in future years rather than travel with conspecifics and be forced to live in areas with which they were not familiar.

The simple view so far presented is that birds explore when young and then perform only calculated and revisiting migrations when adult, restricting themselves to sites with which they are familiar. There is no reason, however, for exploration to be suspended completely during adulthood. Adult humans often visit new places. It is inevitable, however, that most exploration will occur when young. When a bird is born it is familiar with nowhere; by the time it begins to reproduce it needs to be exploiting its environment to maximum efficiency. Nevertheless, environments change and assessments made when younger may not always be, or remain, accurate. There may always be some advantage in further exploration when adult. Using the exploration model, Mead (1979) and Mead and Harrison (1979a, b) suggest that adult Sand Martins/Bank Swallows, *Riparia riparia*, continue to explore, though less so than juveniles, particularly during the post-breeding period. Such exploration should be particularly advantageous for this species because the vertical sandstone cliffs in which they breed are subject to erosion and a site may disappear completely between autumn and spring while the birds are thousands of kilometres away.

Post-breeding migration is a feature of many bird species and perhaps it is often a process of exploration, or at least revisiting, to assess whether the site in which breeding occurred this year really was the best place to breed. Alternatively, it could be a calculated migration by the bird to a place discovered the previous year that, although not best for breeding may be best for feeding or moulting. Moult migrations are conspicuous in many water birds but perhaps they also occur, at the individual level, in many other species.

We concluded that human pastoral nomads build up their familiar area using the same senses and mechanisms as those used by the more sedentary agriculturalists and industrialists. We can now extend this same principle to birds. No feature of the lifetime tracks of long-distance migrants requires senses or mechanisms other than those shown by more sedentary species. The exploration model explains both. The next, and perhaps more difficult question to answer, mainly because of a lack of data, is whether the same is true for fish that migrate long distances. Do Salmon, Cod, Herring, etc. have a sense of location comparable to that seemingly possessed by all other vertebrates?

The life-histories of the various species of Salmon are well known (Fig. 9.7). Eggs are laid in streams or rivers, depending on the species. In two

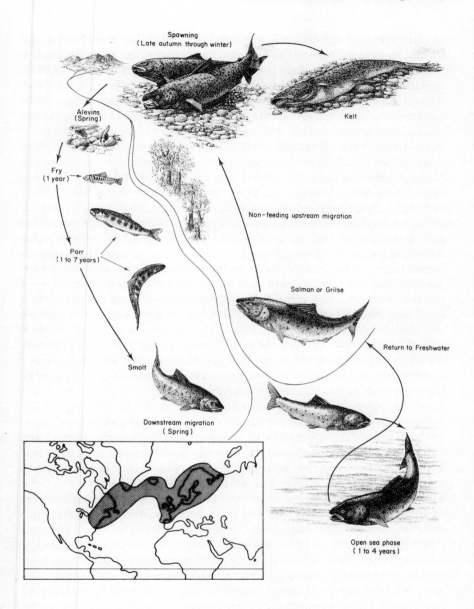

Fig. 9.7 Life history and ontogenetic return migration of the Atlantic Salmon, *Salmo salar*
Death rate after spawning is high, especially among males, but between 20 and 36 per cent of kelts survive to perform downstream migration. The period spent in the sea between spawnings varies from about 4 to 18 months. On average, 3–6 per cent of fish that spawn in a river have spawned previously, though in some short-course rivers in western Scotland and eastern Canada, the proportion may be as high as 34 per cent. A Salmon has been captured that was 13 years old and that had spawned four times.
[From Baker (1978a), partly after Mills]

species of Pacific Salmon, *Oncorhynchus* spp (i.e. Pink, *O.gorbuscha*, and Chum, *O.keta*) the young fish migrate downriver to the sea within a few weeks of hatching. In other Pacific Salmon and in the Atlantic Salmon, *Salmo salar*, one or more years are spent in freshwater before migration to the sea takes place. This migration is performed by a special stage in the animal's development, called the smolt. After one or more years at sea, the adult salmon return to freshwater to spawn, usually in the same stream or river in which they themselves were spawned two or more years earlier. After spawning, all Pacific Salmon die without ever returning to the sea. Many Atlantic Salmon also die but many survive and return to the sea as kelts to migrate upstream and spawn again on future occasions.

We can, for present purposes, ignore the period that the Salmon spends in the sea. Even less is known about this phase of the life-history in Salmon than in oceanic fish such as Herring and Cod. What is known suggests that while at sea Salmon behaviour is similar to that of oceanic fish. If we confine ourselves to the freshwater phases of the Salmon life-history, the critical questions are not difficult to find. Do Salmon 'imprint' on their natal site and then return to it blindly, with no sense of location, simply orientating upcurrent whenever they perceive their natal imprint? Or do they explore for the best spawning site, make assessments, and then respond accordingly? If they explore, do they do so as young fry and parr during their gradual downstream migration, as smolt during their more rapid downstream migration, or as returning adults? Critical data are scarce. We have to make do with what are available.

At first sight, the evidence seems to support the imprinting model. Eighty to ninety per cent of all Salmon that survive to return to freshwater do so to their natal river or stream. The elegant experiments of A.D. Hasler and his colleagues have shown conclusively that chemical cues picked up during early life are of utmost importance in enabling the Salmon to return to its natal site. This is certainly navigation by olfaction (Chapter 10), but is it imprinting? What about those 10−20 per cent of surviving fish that return, not to their natal site, but to some other stream, river or even drainage system? Are they 'strays' as usually considered, individuals in which the navigation system has failed, or are they explorers? Indeed, if spawning elsewhere, are they individuals that, having explored, have judged some other spawning site to be better than their natal site? These 'strays' are the key to answering our question. First, though, consider the movements of the young fish before migrating to the sea.

The fry and parr of most Salmon and Trout are territorial, spacing themselves out in aggressively-defended feeding home ranges (Le Cren 1973, Gee *et al.* 1978). As they grow, they gradually move downstream to deeper and deeper water, small, low RHP individuals travelling further downstream than larger individuals of the same age. During this downstream movement, individuals reject some sites, accept and compete over others, and show an ability to return to territories from which they have been displaced experimentally (Kalleberg 1958). Young Salmon and Trout undoubtedly explore for suitable territories, but are they at the same time exploring with respect to potential spawning sites? They certainly show behaviour that would give them the opportunity to do so. Young salmonids migrate not only downstream but also upstream, leaping small obstacles as they go (Symons 1978). Upstream locations are more likely to be visited in winter while adults are spawning. Young Rainbow Trout, *Salmo gairdneri*, respond to long days by moving downstream and to short days by moving upstream (Northcote 1958). Parr of the Atlantic Salmon may appear in large numbers on the spawning grounds and some of these tiny males may even take part in spawning, swimming down inside the redd cut by the female and fertilising eggs at the bottom of the pit which were not reached by the sperm from the female's fully grown male partner (Jones 1959).

So young Salmon do have the opportunity to assess the potential of different spawning grounds. Moreover, there is indirect evidence that they register the characteristics of life in their particular drainage system. To make the point, though, we have first to consider the timing of upstream migration by adults in different rivers. Inevitably, the time required for the upstream spawning migration varies according to the distance from the sea of suitable spawning grounds. Thus, in the USSR, the Omul, *Coregonus autumnalis*, an anadromous fish like the Salmon, goes 1600 km up the River Lena but only 500 km up the River Jana. In order to arrive at the spawning grounds at the appropriate time of year, it is obvious that anadromous fish have to begin their journeys at very different times in different rivers in order to arrive at the spawning grounds at the best time of year. Even in a limited area such as the British Isles there is much variation in the time that Salmon enter the different rivers. Some rivers have spring runs of Atlantic Salmon, others summer runs, and so on (Mills 1971). To return to the main point, if Salmon fry or parr are transferred from one stream to another, not only do the fish, when adult, return to the stream to which they were transferred rather than the stream into which they were spawned, but most often they return at a time of year appropriate to their adopted stream (Mills 1971). Evidently at some time during the fry–smolt period, the young salmonid learns the best time to return to the drainage system in order to arrive at the spawning ground at the appropriate season. The most likely way for such timing to be learned would seem to be by observation of returning adults, probably during the time the young fish are in freshwater.

Although young Salmon seem to have the opportunity for exploration, the scope must be far greater when the fish returns as an adult, if only because of its much greater mobility. This brings us back to the key question: what are the so-called 'strays'?

There is some evidence that Salmon do not simply return directly to their natal site and then spawn. Even when they have relocated their natal stream, they may visit some other stream before spawning. Sockeye Salmon, *O. nerka*, were captured at the mouth of three tributaries and the outlet stream of Brook's Lake, Alaska, on the shoulder of the Aleutian Peninsula (Hartman and Raleigh 1964). The fish were then divided into four groups and released in each of these streams. As usual, about 75 per cent of all fish released in some other stream were later recaptured back in their original stretch of water. About 1 per cent remained in the stream in which they were released and about 2 per cent were recaptured in some other stream. These may or may not have been 'lost'. Of particular interest, however, was the fact that nearly 5 per cent of all fish released back into the stream in which they were captured were nevertheless recaptured later in some other stream. Such results would be expected if the returning salmon were carrying out exploration, or revisiting sites explored when young.

There is also evidence that Salmon are judging the suitability of rivers and streams as they make their way back toward their natal area and that 'strays' are often individuals acting upon the judgements made. This is indicated by comparing the response of fish to streams that are barren because they are actively avoided with the response to streams that are barren only because of a physical obstacle, such as a dam. Attempts to re-stock rivers that do not support a Salmon run or to replenish poor runs have invariably been disappointing (Harden Jones 1968). Few fish return. Where data exist, this failure is due not to high mortality but probably to avoidance of the stream concerned. In one experiment on Pink Salmon off the Pacific coast of Canada, only 7 adults returned from 1.4 million individuals that survived transplantation to migrate downstream. This rate of return (0.0005 per cent) compares with the usual rate of return of between 0.04 and 7.08 per cent. At the time these Salmon should have been returning to their new home, 40 (i.e 0.03 per cent) of them, all adult, were recaptured in the vicinity of the Fraser River, 500 km to the south of the stream in question. The indication is that there is no run of Salmon to the re-stocked stream because it was in some way unsuitable. Moreover, this seems to have been recognised by the young fish that hatched from the transplanted eggs during the time they spent in their adopted river. In consequence, when adult they sought some other drainage system in which to spawn.

Contrast this with what happens when a dam is constructed and then removed a few years later from a stream that supports a good Salmon run. The classic example derives from the experiments by White (1936) on the Atlantic Salmon in Apple River, a small stream in Cumberland County, Nova Scotia. Both East and West branches of Apple River used to have Salmon runs but in the 1870s a dam was built across the East branch and the run of Salmon to that branch stopped. The dam was removed in the 1920s but until 1931 Salmon had not returned to spawn, though they continued to do so in the West branch. In July 1932, 25 000 fry were planted in the East Apple River. A month or so later, adult Salmon entered the East branch for the first time and spawned. Other Salmon runs occurred in the autumns of 1933 and 1934. Salmon from the transplanted fry returned to Apple River for the first time in the autumn of 1935. Of these, 94 per cent spawned in the East branch and about 6 per cent in the West branch. If we assume that one of the factors a Salmon would take into account in assessing suitability would be the presence of fry, evidence of successful spawning by conspecifics the previous year, the addition of fry to an otherwise suitable stream would be all that was necessary for the East branch to be assessed as suitable for spawning. Whether or not this interpretation is accepted, the Apple River experiment suggests that Salmon do not return blindly upstream to their natal site, homing in on an irreversibly imprinted chemical pattern. Some perception and assessment of their environment is taking place. If salmon really do assess and compare potential spawning sites during their migration, this fact has to be borne in mind as we attempt to re-stock all those rivers in the process of recovering from decades of pollution. We cannot, as the ethological model would suggest, simply ladel hatchery-raised fry into a river and, as long as they manage to survive, expect them to imprint and return in large numbers if they are likely during their travels to encounter somewhere better.

There seems no reason, then, to consider the lifetime track of Salmon, at least while in freshwater, to be the result of mechanisms any less cerebral than those shown by all of the other vertebrates considered in the last few chapters. The final bastion of any hopes that among vertebrates there will be found animals without a human-type sense of location lies with oceanic fish. To judge from the gulf between the classical view and the view encapsulated within the exploration model, the bastion seems fairly sturdy. Consider the classical view. Past attempts to construct migration circuits based on animals at the mercy of the currents, with no sense of location except perhaps a tenuous olfactory thread leading back to some limited home site, bears testimony to the previous ethological view of the individual fish; a non-sentient creature living in a virtual sensory void. Is it possible that the data that could have been considered to support such a view really contained evidence of a different creature; a sentient individual, that explores, builds up a map of its environment, has a human-type sense of location, and makes judgements concerning which sites are good and which are bad? Such a creature can be found, though data are few, and critical data do not exist.

It seems clear enough that most migration circuits of oceanic fish are predominantly downcurrent rather than upcurrent (Harden Jones 1968). It also seems clear that some migration circuits involve at least short periods of migration against or across the current. Examples are: (1) shoals of Herring, *Clupea harengus*, which, on their way in January to spawning grounds off the Norway coast, cross the North Atlantic current that flows to the north-east (Fig. 9.8); and (2) other shoals in the southern North Sea which migrate against the current on their way south down the east coast of the Southern

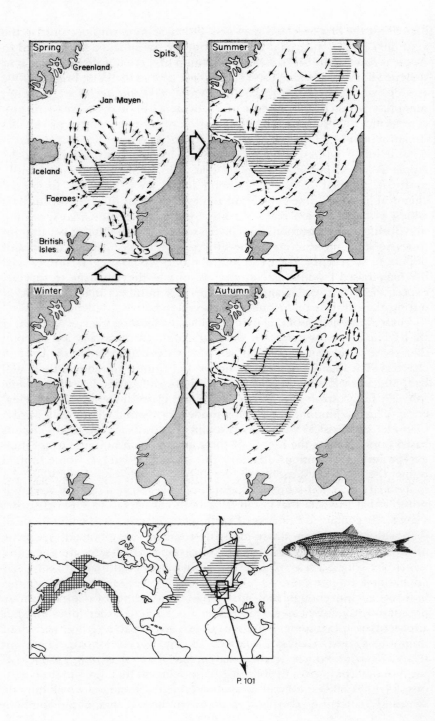

P. 101

Fig. 9.8 Annual migration cycle of adults of the spring-spawning Norwegian demes of Herring, *Clupea harengus*
Horizontal hatching, distribution; dashed line, limits of distribution during the previous season; arrows, surface ocean currents; solid black area in spring diagram, spawning grounds. In order to reach the spawning grounds from their winter distribution, the Herring have to cross, or dive beneath a north-easterly flowing surface current. Otherwise, most of the migration circuit is with the current and is geared to follow biological spring as it spreads north-west.
[From Baker (1978a), after Harden Jones]

Bight (Fig. 9.9c). Any migration model has to be flexible enough, therefore, to allow fish to adopt a direction relative to the current that allows adaptation to local conditions, even if their preference is to migrate downcurrent wherever possible.

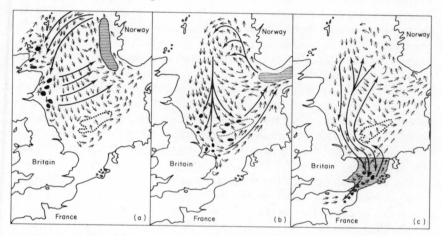

Fig. 9.9 The annual migration circuits of three spawning demes of Herring, *Clupea harengus*, in the North Sea
Solid black, spawning areas; horizontal hatching, wintering areas. Deme (a) spawns in summer and autumn; deme (b) in autumn; and deme (c) in winter. The dotted outline in the middle of the North Sea shows the approximate position of the Dogger sandbank. The long arrows indicate probable migration routes. Short arrows indicate surface ocean currents.
[From Baker (1978a), after Harden Jones]

Many of the young of Norwegian spring-spawning Herring spend the first year of their life in the fjord or coastal waters between the spawning areas (Fig. 9.8) and the coast of northern Norway. The precise site depends on how far they drift downcurrent during their first months of life. Others develop in more open waters. Immature Herring carry out an annual inshore−offshore return migration cycle, moving inshore for the summer and offshore for the winter. This cycle is repeated each year until the young fish reach a certain critical length, whereupon they leave coastal waters and spend one or two years at sea. If there is to to be a period of oceanic exploration during their life-history this will be it, and indeed during this period they are much wider ranging than adults, being found all over the Norwegian Sea. An annual migration circuit (Fig. 9.8) only crystallises out of the area travelled as immatures when, in the winter of their fourth to seventh year of life, they join a spawning shoal. These coastal and then pelagic phases of immaturity give ample opportunity for exploration and assessment of potential spawning and feeding habitats.

In the North Sea, Herring metamorphose at a length of 3—5 cm and then move offshore to deeper water when 9—10 cm long, particularly during the spring of their first or second year. These young Herring become most numerous in the southern North Sea, particularly in the area south and east of the Dogger Bank (Fig. 9.9). In the autumn the fish spread out east, north and west and by the following spring most are north of the Dogger Bank, migrating further and further to the north as spring and summer progress. Eventually, they are spread out all over the middle North Sea. Those that do not recruit to spawning groups during their third or fourth years probably feed in the western part of the middle and northern North Sea and winter more towards the east. Annual migration circuits do not crystallise out from this much more diffuse area travelled during immaturity until the Herring recruit to spawning shoals. Again, therefore, the movements of immatures are just as would be expected if they were exploring rather than immediately settling down to a pattern of downcurrent drift. There is no evidence either way, however, to indicate whether habitat assessment and comparison is taking place. By contrast, there is just a hint that this may occur when we consider recent changes in the migration pattern of Cod, *Gadus morhua*.

The twentieth century has witnessed a northward shift in the geographical distribution of many boreal marine animals. The Tunny, *Thunnus thynnus*, is one such species. The Cod is another. The reasons are unknown, but the shift reflects a general warming of North Atlantic waters. Whatever the reasons, it seems likely that, at the beginning of the twentieth century, there were no Cod spawning on the banks off West Greenland (Harden Jones 1968). From 1917 onwards, however, there has been a great increase in the abundance of the species in the region. The majority of the initial colonists in 1917 seem to have been 5-year-olds, removal migrants from the Iceland-spawning deme of Cod that remained to spawn off Greenland instead of returning to Iceland (Fig. 9.10). By the 1960s, the West Greenland Cod seemed to consist of two incipient groups. The southernmost fish seemed to continue to return to Iceland to spawn whereas the northernmost fish seemed to be members of a newly-formed West Greenland deme with its own migration circuit (Fig. 9.10). This latter circuit seems well suited to local conditions but shares only the grossest of similarities with the migration circuit of its parent deme. Such formation of a new migration circuit is easily conceived if we accept crystallisation from out of a familiar area built up by a process of exploratory migration and habitat assessment during immaturity.

Suppose that, after metamorphosis, immature fish do explore. Suppose further that the immatures of even oceanic fish, such as Herring and Cod,

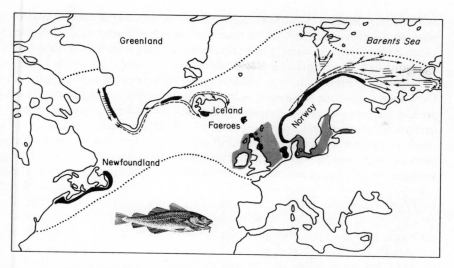

Fig. 9.10 Migration circuits of three demes of the Cod, *Gadus morhua*
Dotted lines, approximate limits to distribution of Cod; solid black, major Cod spawning
areas; stippling, areas where spawning occurs but with small numbers; horizontal lines,
wintering areas. Continuous arrows indicate migrations of mature fish whereas dashed
arrows indicate the drift migrations of eggs and larvae from the spawning grounds to the
nursery grounds.
[From Baker (1978a), after Harden Jones]

are mobile enough to build up a wide-ranging map of the ocean that they
then store within their spatial memory. Suppose, finally, that the migration
circuit eventually adopted is based on the habitat assessments made during
this exploratory phase, biased no doubt by perception of the activities of
adults during the young's own movements through the seas. In the final
year before sexual maturity, for example, the young Cod attaches itself to a
shoal and performs a complete year's migration circuit before its own first
spawning. Such a 'trial' circuit could represent the final stages of
exploration and habitat assessment. Suppose all this, which, apart from its
scale, is asking no more than seems to hold for other vertebrates; how do
we then accommodate the movements of the animal before metamorphosis when, as a young larva or even as an egg, it may drift over
distances of hundreds of kilometres with the current?

Pacific Herring, *Clupea pallasi*, lay their eggs on seaweed between tide
marks and Atlantic Herring, *C. harengus*, lay their's on shingle or gravel
beds in depths from 40 to 200 m. The eggs are fixed but the larvae that
hatch from them are pelagic and drift with the current. Cod are pelagic
both as eggs and larvae, spending a fortnight or so drifting as eggs and 3 – 6
months as larvae. Cod spawned as eggs off southern Iceland have drifted
with the Irminger Current and arrived off south-west Greenland before
they metamorphose and take to the sea bottom. Others drift round to the
northern coast of Iceland. Larvae from spawnings off the Norway coast

drift to the Barents Sea (Fig. 9.10). Both eggs and larvae of plaice, *Pleuronectes platessa*, drift with the current; eggs hatch within 2−3 weeks and metamorphosis takes place between 5 and 8 weeks later. The relationship between spawning and the nursery grounds where the young fish eventually take to the bottom and move inshore is shown in Fig. 9.11 for 3 demes of North Sea Plaice. Rate of migration of eggs and larvae from the Southern Bight spawning grounds is about 3.8 km/d. The total distance between spawning and nursery grounds is about 260 km.

This seems to be the final test. Can these young larvae really have a sense of location during the drifting phase and when they become more mobile and start to control their own movements relate their movements to the

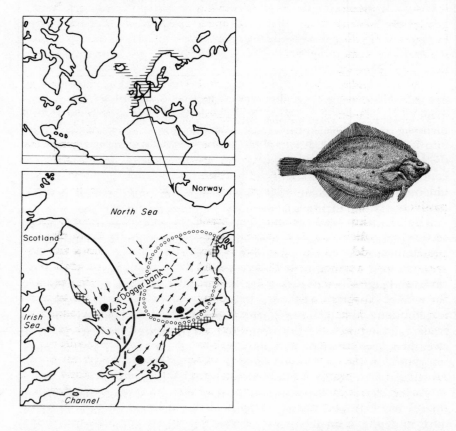

Fig. 9.11 Ontogenetic and seasonal return migrations of the Plaice, *Pleuronectes platessa* During the November to May spawning season, Plaice eggs can be found over a wide area in the southern North Sea, but there are major concentrations such as those shown by large solid black dots in the bottom map. The top map shows geographical distribution. The approximate nursery areas, to which the eggs and larvae are carried by the current, are shown by cross-hatching. Solid and dashed lines and row of circles indicate the limits of subsequent recaptures of mature Plaice tagged on the three spawning grounds indicated. [From Baker (1978a), after Harden Jones]

route they travelled as larvae? Even if they can, what about the eggs? Can an egg have a sense of location?

The first thing we need to know is whether a fish, when it spawns for the first time, shows evidence of being able to return to its natal spawning site. Such evidence is difficult to obtain because eggs and larvae cannot be tagged in the same way as larger fish. Nevertheless, several lines of circumstantial evidence (see critical review by Harden Jones 1968), such as analysis of body measurements and scales, suggest that at least Norwegian Herring and Cod and North Sea Plaice return to spawn for the first time to the general area of their natal site. However, the precision of return may not be high, the fish returning to only the general area, not the precise site. It would be interesting to know if Herring, with their fixed eggs, have a greater precision of return than Cod and Plaice with their drifting eggs. Perhaps it is significant, also, that Herring recruit directly to their spawning shoals. Cod, on the other hand, join spawning shoals a year or more before they themselves spawn.

We can find no real clue in our search to decide whether a floating egg, or rather the developing embryo, has sufficient awareness of its environment to gain a sense of location. The apparent ability of oceanic fish to find their way back to the general area in which they were spawned or hatched into larvae, however, does indicate that a sense of location is gained by the larva. As it drifts with the current, enough information is collected concerning location and change of position for the same individual, when older and exploratory, to relate its travels to the route taken as a small, passively drifting larva.

The fact that adult oceanic fish travel in migration circuits large sections of which are downcurrent in no way detracts from the exploration model and the view that as adults they live within a familiar area and have a strong sense of location. In the same way that adult birds take advantage of winds to travel in their standard direction, often waiting for winds of an appropriate direction, so too should we expect fish to assess not only which are the best areas but also which is the most economical route by which to travel between them. If it is possible for a fish to spend most of its year travelling downcurrent between habitats we should not be surprised (on the exploration model) to find that they elect to do so. Equally, if two highly suitable habitats can only be incorporated into a migration circuit by travelling across or against a current, the exploration model would expect that to happen also. Finally, if an area of ocean adjacent to a traditional migration circuit for some reason becomes suitable for occupation and is discovered by explorers, then the exploration model would expect that area to be colonised and the fish to work out a new and appropriate migration circuit.

Nobody could pretend that the evidence on the movements of oceanic fish provides overwhelming proof for the exploration model, though we can, I think, dismiss the notion of passive downcurrent drift in a sensory void. Doubtless an ethologist could work out a model based on reflex response to immediate stimuli (as Harden Jones has done so ingeni-ously)that also fits all the facts, few as they are. Which view we adopt until critical data are available depends on our premise—on our paradigm.

So far, and quite rightly, the paradigm for animal movements has been the one that requires least from the animal in terms of cerebral activity. The entire field of optimal foraging (see review by Krebs 1978) began with the paradigm of random movement. The first step was to show that in the real world animals did better than if they travelled randomly through the environment. For a time, random movement was the model for the migration of oceanic fish (Saila and Shappy 1963) and the homing of birds (Wilkinson 1952). Again it was necessary (see Harden Jones 1968 and Matthews 1968) to show that in the real world animals did better than if they were moving randomly. Throughout the age of ethology the basic assumption for animal migration in general was that it was based on imprinting and programmed orientation and navigation; that other animals did not use a cerebral sense of location of the human type; that they dispersed instead of explored; and so on.

Any science proceeds by a process of setting up hypotheses, testing them, and if necessary rejecting or modifying them in favour of some other hypothesis. I advocate that now is the time, as the study of animal behaviour moves into the age of Behavioural Ecology, to change our hypothesis concerning animal movements. Enough evidence is available for it now to be accepted that vertebrates and a wide range of invertebrates live within a familiar area; that this familiar area is built up by a process of exploratory migration and habitat assessment; that there is no difference between the mechanisms used and those used by humans. This should be the new hypothesis.

Let us accept that the movements of young Lesser Black-Backed Gulls, *Larus fuscus* are exploratory; that during the first four months of their life they associate with, or at least observe, the movements of parents or other adults; and that their migration pattern when adult is based on habitat assessments and rankings carried out when young. Over the past few decades, the winter ranges ranked most highly have evidently been further and further to the north. In the northern hemisphere, we normally think of birds breeding in the north and then flying south for winter. How then is it possible for young *L. fuscus* to explore northern areas in winter in order to judge how they compare with wintering grounds further to the south? Figure 9.12 plots the distribution of *L.fuscus* for each month or so of the year during the first three years of life and when adult. The pattern that emerges is as follows. When first-year birds begin exploration independently of adults, they use as a basis a basic familiar area spanning the entire distance from the breeding grounds to North-West Africa. During their first winter they explore intensively the southernmost part of this area, perhaps extending familiarity even further to the south than they travelled with

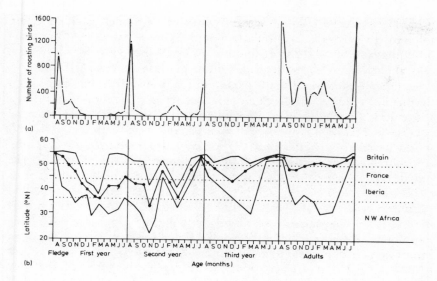

Fig. 9.12 Variation in the distribution of Lesser-Black-Backed Gulls, *Larus fuscus*, with age and time of year

(a) Counts of *L. fuscus* at a night roost on Audenshaw Reservoir, Manchester, July 1977 to July 1978. Third year birds were lumped with adults during counting.

(b) Mean (solid dot) and northern and southern limits (solid lines) to the latitudinal distribution of *L. fuscus* ringed as pulli on Walney Island, England, and recaptured as dead birds, 1969–1975.

[Simplified from Baker (1980a)]

adults. In the following spring and summer they return briefly to the north to visit (and investigate) the breeding grounds but return to the south in advance of the adults and the year's new crop of young. Having returned to the south, however, they then migrate back to the north during their second winter. This northward winter migration by second-year birds is shown not only by ringing recoveries but also by the appearance of such birds among winter roosting flocks (Fig. 9.12a). In spring, the birds migrate back to the south before returning north once more, again to visit the breeding grounds as non-breeding birds. During their third year, the adult migration pattern begins to emerge.

The indication is that these birds spend their first winter exploring and extending the southern part of the area visited during the first four months of life with adults, and their second winter exploring and extending the northern part of the area. This pattern has remained constant, at least since 1962 (Baker 1980a).

The Barren-Ground Caribou, *Rangifer tarandus*, of Canada performs an annual to-and-fro migration between forested winter grounds and summer parturition and feeding grounds on the tundra to the north or

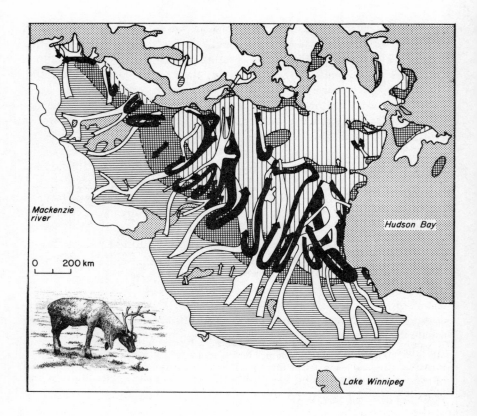

Fig. 9.13 Seasonal return migration of the Barren-Ground Caribou, *Rangifer tarandus* in Canada
White arrows, spring migration; black arrows, autumn migrations. By late September, during autumn migration, the herds arrive at and move along the tree-line or perhaps even back onto the tundra again, as shown. Copulation occurs in late October and early November, usually in the region of the tree-line, before the herds move into the forests for the winter. Horizontal hatching, winter range; vertical hatching, summer range, on the tundra. Caribou start the spring migration sometime between February and April and reach the tree line on the way north in the first week of May. From the tree-line to the nearest calving grounds is 400–500 km. Width of arrows indicate relative number of animals involved.
[From Baker (1978a), after Banfield and Kelsall]

north-east (Fig. 9.13). Young Caribou undoubtedly establish a basic familiar area during their first year of life by association with their mother, and data concerning any independent exploration that occurs in later years are few. There are two points of interest. Firstly, following on from the account of the Lesser Black-Back, we can ask whether the young Caribou takes for granted the fact that the best winter home ranges are to be found in the forests as shown them by their mother or do they judge this for themselves in the way that the Lesser Black-Backed Gulls do. There are frequent observations of small herds of Caribou on the tundra in winter, eking out a living on wind-swept slopes where the snow never becomes deep. Could these be explorers, assessing whether the tundra really is unsuitable as a winter range? Secondly, there is for me an interesting question mark over the dispersal (using the term in its legitimate sense) of the great herds that occurs each August. During this month, the vast majority of individuals are scattered across the tundra in small groups of three or less. Could this be an annual period of exploration? At a rate of, say, 50 km/day, a Caribou could travel over 1000 km during this period. At such a rate, a Caribou could, before it was many years of age, be familiar with a large part of the Canadian tundra.

Why should a Caribou go to such lengths to build up such a vast familiar area? What possible advantage could there be to warrant such expenditure of time and energy? The answer could be that every few years Caribou have to abandon their current migration circuit and set up a new one elsewhere. Such removal migrations are not easy to detect but their occurrence can be deduced from the fact that over the years the greatest density of Caribou oscillates between the western and eastern parts of the Canadian tundra. Kelsall (1968) presents two examples of herds abandoning their current migration circuit and migrating at right angles to the usual north-east/south-west axis before setting up a new migration circuit elsewhere (Fig. 9.14). On both of these occasions the direction of the removal migration was toward areas of reduced deme density. Indeed, one of the migrations led to the recolonisation of an area devoid of herds for 15 years. Such shifts seem to be necessary because on the tundra the foodplants of the Caribou (lichens, mosses, and the leaves of a wide variety of other plants) are slow to recover from grazing. So, again we can ask the question: how do Caribou decide when and where to go to set up their new migration circuit? How has the information been collected if we accept as a premise that such removal migrations are calculated, taking place within a familiar area? Is this the explanation for the August dispersal of the great herds?

We may as well end on a speculative (some would say far-fetched) note. Perhaps August is a time of individual exploration. Perhaps when the herds reassemble, individuals compare information and the herd makes a democratic decision concerning its movements during the coming year; whether to persist with the established circuit or whether at some stage during the coming year to migrate to a new circuit. Far-fetched? If Honey Bees behave in this way, why not Caribou?

The last four chapters have been concerned with exploration. We have seen that all vertebrates and a wide range of invertebrates live within a familiar area. Moreover, we have seen that all such animals build up their familiar area in a fundamentally similar way, by a process of exploratory migration and habitat assessment and ranking. This remains true whether the animal concerned lives in more or less the same place for much of its adult life or whether it is one of the classical migrants, shifting from one seasonal home range to another, perhaps travelling thousands of kilometres. Throughout these chapters, it has been necessary, because of the conceptual legacy from the past, to keep looking over our shoulders at Man, just in case there is something special about the way he builds up his familiar area. We conclude that there is nothing special and that we can afford to be, indeed we must be, anthropomorphic. We must attribute

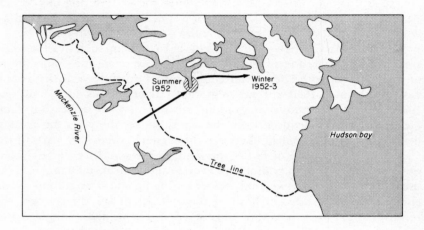

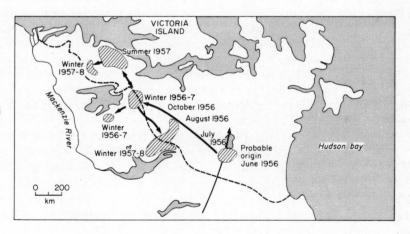

Fig. 9.14 Long-distance removal migration by large herds of Barren-Ground Caribou, *Rangifer tarandus* on the Canadian tundra
On both of the occasions shown, herds abandoned their previous migration circuit and travelled at right angles before establishing a new circuit elsewhere. The migration in the top diagram led to the recolonisation of areas devoid of herds for 15 years.
[From Baker (1978a), after Kelsall. Photo by J. P. Kelsall (courtesy of Canadian Wildlife Service)]

these other animals with a cerebral sense of location quite the equal of that shown by Man.

Only one element of exploration and sense of location has so far been neglected and, when considered, a curious twist is revealed. The missing element is navigation, behaviour that for some strange reason Man has long been prepared to acknowledge as more highly developed in other animals than in himself. In the next chapter, therefore, our approach changes. No longer have we to question Man's superiority; it is his supposed inferiority that is of interest.

10

Navigation: using the sense of location

In the minds of most people, the words 'migration' and 'navigation' are inextricably intertwined. Migrants navigate; navigation is what migrants do. This was the axiom that threaded its way through the age of ethology, despite the paradox of the Homing Pigeon (Chapter 2). Impressed to the point of being overawed by the way birds returned each year to nest in the same site, even after journeys of thousands of kilometres, amazed to the point of disbelief that a salmon can return to spawn in the same stream in which it was itself spawned, Man has for decades studied such animals in search of unknown mechanisms and senses not possessed by humans. Navigation is probably the only field of behaviour in which Man has been prepared to accept that, without the aid of instruments, he is inferior to these other animals. Perhaps inferior is the wrong word; deficient is more apt. Man saw himself as lacking this mystic ability by which birds and other animals returned home with such precision. As recently as 1978 one of the leading research workers in the field of bird navigation proclaimed that: 'We *must* discover a new sensory channel— the ones we now have are not sufficient to explain the animal's behaviour'.

The previous chapters have shown that there is nothing unusual about the behaviour of the classical migrants. No matter whether we think of a human industrialist or a pastoral nomad, a mouse or a Caribou, a House Sparrow or a Swallow, a Giant Tortoise or a Green Turtle, a Goldfish or a Salmon, our picture of the way they build up their familiar area is always the same. If this picture is correct, where does it leave the study of navigation? Does it imply that long-distance migrants do not have a special navigational ability or does it imply that more sedentary animals have navigational ability also? Either way we are freed from the Homing Pigeon paradox. Does it also suggest that all these years Man had no need to feel that he was deficient compared to other vertebrates? These are ambitious questions; we must begin carefully by making certain we all mean the same thing when we talk of 'navigation'.

Some people use the term in a fairly haphazard way to describe any type of directional response that is related in some way to geophysical or celestial cues such as the Earth's magnetic field or the Sun and Stars. Others are more stringent and use it to mean the determination of the direction of a particular destination, but again only if the cues used are geophysical or celestial. Such authors contrast navigation with *pilotage*—the process of finding one's way across familiar terrain by the use of familiar landmarks. The first of these two usages does not allow us to distinguish the behaviour of animals with a sense of location from those with primarily a sense of direction (Chapters 11 and 12). The second usage cannot be upheld because in finding their way around all animals probably use a combination of geophysical and celestial cues and landmarks. In common usage a navigator is somebody who finds the way to a required destination across unfamiliar terrain, no matter by what means. A car passenger who reads a map and looks for signposts is just as much a navigator as is the sailor who uses only a clock and sextant. There seems no reason to use the word in any other sense. An additional advantage is that it allows contrast with *orientation*, i.e. the maintenance of a particular direction irrespective of destination. It is often difficult to tell whether an animal is orientating or navigating. Indeed, the only way is to displace the animal from its present position and see if it compensates for the displacement or continues parallel to its original course. In the first case the animal is navigating; in the second it is orientating. In the first case the animal has a sense of location; in the second it has only a sense of direction (Fig. 10.1).

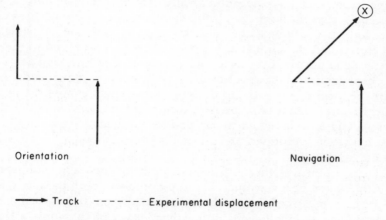

Fig. 10.1 Lateral displacement as a means of distinguishing between orientation and navigation.
[From Baker (1978a)]

Such a distinction may seem simple, but there are pitfalls. Consider again the displaced navigator in Fig. 10.1. Suppose that, upon release, the animal uses its navigation mechanism and decides that home is in a particular direction, say north-east. As long as the animal can then find some environmental cue that allows it to maintain a north-easterly direction it can return home simply by orientating. The homing process, following displacement-release, may involve navigation for only a few seconds followed by a long-period of orientation. To study navigation, therefore, it is vital that the animal is monitored at the time it is making the decision concerning home direction. Previously, it was thought that this was the period immediately following release but we now know that many animals have reached a decision, though not an irreversible one, even before they are released. Any observations after the animal has made its decision may reflect only orientation, not navigation. The interpretation of *homing success*, the proportion of released animals that return home and the speed at which they do so, is in consequence fraught with difficulties.

Continuing the theme of navigation versus orientation, Matthews (1955, 1968) showed that when Mallard, *Anas platyrhynchos*, from Slimbridge were displaced their initial orientation upon release was always to the north-west. This was true whatever the direction of Slimbridge from the release point. It was also true whether or not the birds were released within their familiar area (Matthews and Cook 1977). Matthews called this *nonsense orientation*, an unfortunate term because it implied that there was no sense to the behaviour. Yet sense there has to be. The difficulty is in finding it. Wallraff (1978) has replaced the term by *preferred compass direction* (PCD) and has shown that each loft of pigeons has its own different PCD, though most have a PCD roughly to the west. In the early days of experimentation on navigation, many releases were in only a single direction from the home site. If this direction coincided with the PCD of the animals concerned, the impression given was one of navigation, whereas in fact the animals had no idea where they were and were simply orienting in their PCD. This could be demonstrated by taking the animals an equal distance in the opposite direction from the home site and showing that they still orient strongly in their PCD, which is now the opposite direction from home. As a result of the danger of interpreting orientation in the PCD as navigation, standard practice in navigation experiments is to test animals in at least two, and preferably four, compass quadrants relative to the home site.

There are many other pitfalls in experiments on navigation, some of which cannot be overcome. The most important of these is that the experimenter can only guess where the animal wants to go when released. Experimenters using Pigeons assume their birds wish to return to their loft and assess their performance accordingly. If, in fact, the bird knows of a very good feeding area in the opposite direction and decides to go there instead the experimenter would conclude that it was lost or disoriented. Without knowing precisely where each bird wants to go upon release, the experimenter has to settle for a statistical demonstration of navigation (Batschelet 1965, 1978). As long as the majority of pigeons attempt to return directly to the loft, the statistics can tolerate a few individuals that do not.

The same problem is also found when using homing success as a measure of navigational ability. Apart from the fact that most of the homing process involves orientation, not navigation, homing success also depends on the proportion of animals that decide they still want to return home. Figure 10.2 shows the way that homing success varies according to the distance of release. For all animals, the proportion that return decreases as the distance of release increases. Does this tell us anything about the navigation mechanism? Even if this decrease is in part due to an increasing number of

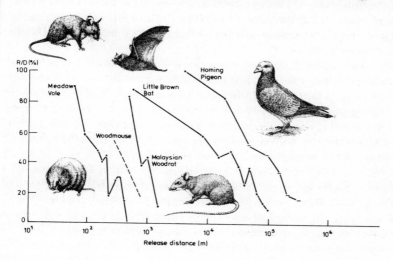

Fig. 10.2 The relationship between homing success and displacement distance in 5 species of vertebrate: Meadow Vole, *Microtus pennsylvanicus*; Wood Mouse, *Apodemus* spp.; Malaysian Woodrat, *Rattus jalorensis*; Little Brown Bat, *Myotis lucifugus*; and Homing Pigeon, *Columba livia* All data are for individuals displaced for the first time. Homing success is measured as per cent return to the home site.
[Redrawn, with simplification, from Bovet (1978), after Robinson and Falls, Bovet and Gogniat, Harrison, Mueller and Wallraff]

animals deciding to stay at the release point rather than try to return to their original home range, it does not necessarily imply that they do not know where they are. On the contrary, it may be precisely because they do know where they are, a long way from their home range, that they decide it is easier or safer to settle down rather than to attempt to return over such a distance.

Proponents of the view that animals are incapable of navigation and that all homing is achieved by random search seized upon this decrease in homing success with distance. Such an effect is just as would be expected if the animals moved randomly and indeed, the random search models of Wilkinson (1952) and Saila and Shappy (1963) were surprisingly good at describing the way that homing success, including the speed of return, changed with increasing distance. Random search models collapse, however, in the face of demonstrations that animals are not randomly oriented upon release and do show evidence of knowing the direction of home.

Historically, there has always been one other important criticism of displacement—release as a means of studying navigation: that it is an artificial situation which places animals in a position they would not encounter in nature, and, in particular, it does not mimic the one situation in which 'true' navigation normally occurs, i.e. during long-distance seasonal migration. This criticism dissolves in the face of the exploration model of animal movements. An exploring animal is faced with precisely the same type of problem to solve as that imposed on it by displacement—release (Baker 1981).

Consider again the process of exploration. An animal enters and travels through unfamiliar terrain, leaving behind it a familiar area to which it will want to return when the exploratory bout is complete. The movement has two major functions: (1) to locate, assess and rank a variety of resource locations; and (2) to incorporate the more useful of these within the familiar area. What does such incorporation involve? Firstly, the animal has to learn some means of recognising the new site on future occasions. Secondly, it has to work out some set of instructions by which it can find its way between the new site and other sites within its familiar area. Put another way, the new site has to be placed within the animal's spatial memory. We already have an image of the spatial memory as its exists for humans (Chapter 5): vague 2-dimensional maps on which familiar locations are arranged relative to each other and to some stable absolute axis such as compass direction; detailed instructions on how to move from one place to another, such instructions being arranged in a hierarchy so that more or less the minimum information and effort of concentration is used for any particular movement between two familiar places.

The features by which a location can be recognised I shall term *landmarks*. Conventionally, landmarks are topographical and are perceived by vision. I shall use the term in a much more general sense and extend it to include any feature that is location-specific, whether the feature is topographical, chemical, or even more abstract such as the direction in which air or water is flowing, no matter by what sense the feature is perceived. It follows that a familiar area consists of memorised associations between the availability of different resources and particular landmarks or sets of landmarks. Information concerning landmarks must be stored in the spatial memory in the form of a map. I shall call this map a *familiar area map* to emphasise that it is built up of familiar landmarks and to contrast it with the *grid maps* postulated by ornithologists and described later in this chapter. The most important point of all to be noted, however, is that the limits of the familiar area map are not set by the limits of the most distant places ever visited but by the limits of the landmarks perceived (and memorised) by an individual during its lifetime.

We shall return to the inevitable involvement of navigation with exploration later. First, however, we should pause and consider the nature of landmarks a little further. In effect anything that can be perceived can be a landmark as long as it is a specific indicator of location. On land, topographical features are the obvious landmarks but equally useful are characteristic smells, sounds, magnetic anomalies, etc. At sea, apart from islands and underwater topography, features such as wave direction, colour and chemical 'signature' (smell and/or taste) of the water, the assemblage of floating plants and animals, and many other features, are all potential landmarks. The organic content of sea water in particular changes from place to place (Aubert *et al.* 1978) and could be a reliable landmark feature for any animal, such as a fish, that can perceive such changes. There is no shortage of landmarks and an animal is limited in what it can use only by its range of senses. Most animals have the senses of sight, taste, touch, smell and hearing. Increasingly, it is being demostrated that animals respond to fluctuations in the Earth's magnetic field, including magnetic anomalies caused by local deposits of iron compounds (Walcott 1978).Temperature variations can be detected and used as landmarks, either for distinguishing north and south sides of a hill or for distinguishing warmer and cooler water currents, for example. Fish have lateral line systems, which detect pressure waves and probably allow perception of short-distance topographic features. Some elasmobranchs also have electric field detectors that can be used, among other things, for detecting the direction of water currents. This is done by perceiving the electric fields generated as ocean currents flow through the Earth's magnetic field (Kalmijn 1978). Fish can detect hydrostatic pressure (Tsvetkov 1969) and can therefore recognise their depth as probably, can most other aquatic animals. Birds, and perhaps other terrestrial animals, can detect changes in barometric pressure (Kreithen and Keeton 1974). Birds are also capable of detecting the infrasounds (0.1 – 10.00 Hz) given off by ocean waves, jetstreams, and major topographical features such as mountain ranges (Kreithen 1978), but it is not yet certain that they can determine the direction of such features, though the evidence becomes more and more promising (Quine and Kreithen 1981). Finally, some mammals, notably cetaceans (e.g. Pilleri 1979) and bats, but also pinnipeds, ground-living insectivores, and rodents, are able to pick out and recognise landmarks using echolocation: the emission of directional sounds followed by perception of the pattern of echoes that are given off by surrounding environmental features (see review by Simmons 1977). It has been suggested that the call notes of nocturnal birds may provide information concerning the terrain over which they are flying by a crude form of echolocation (Griffin and Buchler 1978).

However, we are doing no more than listing the sensory capabilities of animals. Some students of navigation still consider that it is in the search for new sensory capabilities that the answer to the 'mystery' of navigation will be found. In my view, we probably know all the more important sensory channels used by navigating animals. The answer now lies in unravelling the way the animals concerned make use of the senses and landmarks available to them. Nevertheless, there is one important point that emerges from a consideration of animal senses, and that is the distance over which different senses allow landmarks to be recognised. The difference between the familiar area and the familiar area map is the distance from which landmarks can be perceived from the periphery of the familiar area. The distance depends on the senses used by an animal to detect the landmarks. Imagine an animal that has never moved, so that its familiar area is confined to a single point. If the animal is terrestrial and uses vision, it will have a familiar area map that (as long as its view is open and not impeded by vegetation etc.) is much larger than if it uses echolocation, which in turn will give a larger familiar area than if touch only were used. If

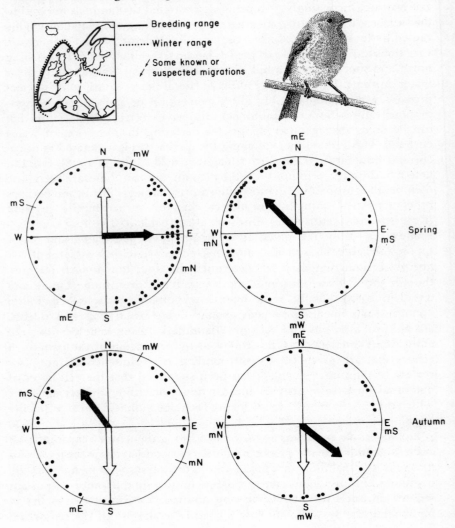

smell is used (and/or taste if the animal is aquatic), the familiar area map will be larger than if vision were used (but perhaps only in one direction, depending on how unidirectional are the air or water currents that carry chemical 'signatures' to the animal). Detection of barometric and hydro-static pressure and of magnetic anomalies would give a relatively small familiar area map, little bigger than the single point that is the familiar area of our imaginary animal. The use of infra-sound could perhaps lead to the largest familiar area map of all, sources of such sound being detectable over thousands of kilometres (Kreithen 1978). We await field confirmation, however, that animals do detect such sources.

Even without moving, therefore, an animal can have a very large familiar area map (the size depending on the senses used) on which the directions of significant landmarks are related to the site occupied. We are

Fig. 10.3 The effect of an experimental manipulation of the magnetic field on the orientation direction of the European Robin, *Erithacus rubecula*
Experiments were performed in circular orientation cages during the periods of spring and autumn migration. Each solid dot indicates the mean direction of hop of a caged individual during migratory restlessness (see chapter 14). The solid arrow shows the mean preferred direction and the open arrow shows the migration direction of free-living conspecifics. N, S, E and W are, respectively, geographical north, south, east and west whereas mN, mS, mE and mW are their experimentally produced magnetic counterparts.
[From Baker (1978a), after Wiltschko. Photo by Ronald Thompson (courtesy of Frank W Lane)]

almost in a position to consider what happens when an animal begins to explore, but not quite. We concluded for humans that the familiar area map was stored in the spatial memory in such a way that not only were familiar landmarks related to each other but they were also related to some stable absolute axis. This is also implicit in the image of the familiar area map that we have just conceived for an animal that has not yet moved. The full significance of this feature of the familiar area map does not emerge until later, but first we ought to consider what range of stable directional axes are available to animals.

Currents of air and water, at least in some areas and some types of habitat (e.g. streams in which water always flows from upstream to downstream), provide a relatively stable, directional axis by which some animals could orientate their familiar area map. In 3-dimensional habitats, such as the ocean, gravity and gradients of hydrostatic pressure and light intensity offer an up–down axis that can also be used in this way. For most animals, however, whatever their habitat, the most reliable and stable directional axis is a compass, a feature of the environment that relates to what humans conceive of as north, south, east and west.

If we confine ourselves to compass cues that we know animals can detect, ignoring potential but as yet unproven possibilities such as Coriolis Force (see Matthews 1968 for discussion), and leave out cues such as air and water currents just described, which in places can act as compass cues, then we only have to consider two types: the Earth's magnetic field and celestial cues such as the position of the Sun, Moon and Stars. Decades of controversy preceded final acceptance of the fact that birds could detect compass direction from the Earth's magnetic field. Now many animals, including mice (Mather and Baker 1981), birds (e.g. Schmidt-Koenig 1979, Keeton 1980, Walcott 1980, Gould 1980; Fig. 10.3), amphibians (Phillips and Adler 1978), fish (Kalmijn 1978, Quinn 1980) and a wide range of invertebrates, including Honey Bees (Lindauer and Martin 1972) and moths (Baker and Mather 1982), are known to have access to a geomagnetic compass. Indeed, the way things are going the final search may be to find an animal that has not such access. The site of the sense organ on which the magnetic compass sense depends has still not been confirmed for any animal. Walcott, Gould and Kirschvink (1979) discovered a region in the head of Homing Pigeons that shows every indication of being a sense organ suitable for detecting such a compass. Details are few but the organ lies between the brain and the skull and consists of deposits of magnetite (but see Presti and Pettigrew 1980). Gould and his colleagues have found an equivalent organ, again containing magnetite, at the front of the abdomen of Honey Bees. Models of the way magnetite could be used to detect direction and location become increasingly sophisticated (Gould 1980, Kirschvink and Gould 1981), though not everyone is convinced that the magnetic sense organ has to involve magnetite, and alternative theories do exist (e.g. Leask 1978). Although birds and other animals have access to a geomagnetic compass, birds at least are reluctant to use it. Homing Pigeons had to be forced to do so by training them always under overcast skies (Keeton 1972a,b). They much prefer to use celestial cues and most, if released in fog or under overcast skies will, unless specially trained, settle down and wait for the skies to clear enough for the Sun to become visible, rather than try to use their geomagnetic compass. The reluctance of birds and others to use their geomagnetic compass if some other is available is perhaps understandable. A magnetic compass is subject to relatively unpredictable fluctuations during the course of a day and from day to day. It is also likely to fail in the vicinity of geographically local magnetic anomalies (Walcott 1978). If celestial compass cues are present, they are much more reliable.

Mammals, birds (Fig. 10.4), reptiles, amphibians, fish, insects and other invertebrates have all been shown to refer to a Sun compass. The position of the Sun's disc may be observed either directly or indirectly from the pattern of polarised light. Bees and some vertebrates can detect the Sun's position through cloud too thick for the human to do so. They can do this because they are receptive to wave-lengths further into the ultra-violet region of the spectrum. Such wave-lengths penetrate cloud better than others (see papers in Schmidt-Koenig and Keeton 1978 on perception of polarised and UV light).

In order accurately to use the Sun as a compass, an animal has to learn to

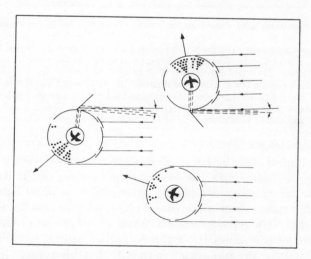

Fig. 10.4 The effect of an experimental manipulation of the apparent position of the Sun on the orientation direction of the Starling, *Sturnus vulgaris*
Starlings were placed in a circular cage, in the centre of a circular room. When exposed to direct sunlight (the room possessing six windows), the Starlings showed preferential compass orientation to the north-west. When the apparent position of the Sun was shifted by a combination of shutters and mirrors, the Starlings changed their compass orientation by an amount equal to the apparent shift of the Sun.
[From Baker (1978a) after Kramer]

compensate for the fact that the Sun moves across the sky during the course of the day. Bees, for example, during the first few days out of their hive, cannot compensate for the Sun's shift. Compensation is learned by noting during the day the Sun's shift relative to familiar landmarks (Frisch 1967) or the Earth's magnetic field (see Schmidt-Koenig 1979). Once the shift has been observed for part of the day, the bee is able to extrapolate the Sun's position at any other time. Bees transported from a place in the Northern Hemisphere, at which the Sun moves clockwise across the sky, to a place in the Southern Hemisphere, where the Sun moves anticlockwise, are at first disoriented. After a minimum exposure of 500 flights over 5 afternoons, however, the bees are found to have an accurate compass based on the newly learned pattern of shift (Lindauer 1967).

That an animal is compensating for the movement of the Sun across the sky can be shown by shifting the animal's clock (Fig. 10.5). Most animals

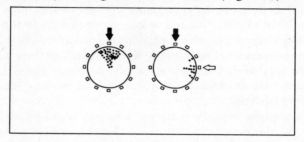

Fig. 10.5 Demonstration of time-compensated Sun orientation by rephasing the biological clock
Starlings were trained under a sunlit sky to feed from a pot in a fixed compass direction as indicated by the black arrow. When, as in the right-hand figure, the birds are exposed for a period to an artificial day/night regime that is 6 h behind the normal day, and then tested as before, they feed from the pot in a compass direction 90° clockwise from that to which they were trained. This experiment demonstrates: (1) that the birds have access to a Sun compass; (2) that they compensate for the Sun's shift across the sky during the day; and (3) that the timing of the light/dark interphase is used as a phase setter for the clock used in such compensation.
[From Baker (1978a), after Hoffmann]

set their biological clock by reference to some consistent event in the daily cycle, often the light−dark transition in the evening or the dark−light transition in the morning. By placing an animal in artificial lighting conditions so that these events occur out of phase with the outside world, the animal's clock can be rephased and set to run a fixed amount of time fast or slow. When such an animal is returned to the natural environment, it should misinterpret the Sun's position by a predictable amount (Fig. 10.5). The stable directional axes of its familiar area map will have shifted.

Few animals have yet been demostrated to have a time-compensated Moon compass, to which the same considerations apply. Among birds only Mallard, *Anas platyrhynchos*, have been shown to have such a compass (Matthews 1973) and among invertebrates only sandhoppers (see review by Enright 1978). So few examples, however, almost certainly reflects a lack of attention from zoologists rather than a lack of use of the Moon by animals. Doubtless all animals that live within a familiar area and move around at night will be found to have a time-compensated Moon compass.

Use of the night-sky to obtain compass information when the Moon is below the horizon has been demonstrated for birds, small mammals, and amphibians. Only for birds, however, has the nature of the compass been demonstrated, and for these there seems to be at least two, perhaps complementary, systems. Emlen (1972) showed that Indigo Buntings, *Passerina cyanea*, use the axis of rotation of the night sky as a reference point. This is equivalent to humans in the Northern Hemisphere using Polaris, the pole star, to indicate north. Not that Polaris, itself, has any inbuilt significance to the birds. When raised in a planetarium they are quite prepared to accept any star within the natural or a completely fictional sky as their reference point as long as it is sited at the axis of rotation; the hub around which the entire sky appears to revolve during the course of a night.

Birds do not need to see Polaris to know its position. In the same way that humans can tell roughly the position of the pole star so long as they can see the constellation of The Plough, Ursa Major, so too can Emlen's Buntings, except that they do not necessarily use The Plough. By blocking out parts of the planetarium sky, Emlen was able to show that different individuals used different star patterns within 15° of the pole star by which to fix the latter's position. As long as each bird could see its personally learned star pattern on a particular night then it knew the location of the pole star and hence had access to a star compass. The way this came about was as follows. During their first summer, before they would normally begin their first autumn migration, the birds noted the movement of the night sky, located the axis of rotation, and learned how to recognise its position relative to nearby star patterns. In Emlen's experiments a revolving sky during this early period was crucial. Without it, the Bunting's would not locate the axis of rotation. Once they had learned how to locate the axis, however, they could recognise North without a rotating sky. In experiments on the Eurasian Robin, *Erithacus rubecula*, Wiltschko and Wiltschko (1976) found that even with a stationary sky night-migrating birds could learn to use a star compass. The birds did this by relating star patterns to the Earth's magnetic field.

Clearly, a great deal more basic research needs to be done before we shall have a complete picture of the range of landmarks and compasses that are available to, and used by, all the animals that live within a familiar area. Until such time, our appreciation of the full content of their familiar area maps can only be partial. Indeed, formal research into the existence and structure of the spatial memory is at a very early stage (Olton 1977, O'Keefe and Nadel 1979). The study of navigation cannot really wait for all this information to be gathered and formalised, nor is it necessary for it to do so. It is probably sufficient to accept that animals do have a familiar area map, stored within a spatial memory; that it is made up of familiar landmarks arranged relative to each other and relative to some stable absolute axis such as compass direction or its substitute; and that it is integrated with detailed instructions on how to get from one place to another. The physical nature of the landmarks and the compass cues are mere details except insofar as they influence on the one hand the distance that the map extends beyond the familiar area and on the other hand the range of conditions under which the animal can find its way from one place to another.

As discussed for humans (Chapter 5), ambient conditions influence which of the memorised landmarks and instructions are used in finding the way from one place to another. This information is arranged in a hierarchy within the memory. The structure of the hierarchy should be such that the animal, in trying to solve any particular navigational problem, should always use the most efficient and economical subset of all the information available. This is the *least navigation* hypothesis (Baker 1978a), which we have already used, in effect, to account for the preference shown by birds for using celestial information in preference to geomagnetic information.

At last we are in a position to return to the relationship between exploration and navigation with which we began. Consider an exploratory migrant that finds a new site suitable enough to be incorporated within the familiar area. The new site has to be 'placed' on the familiar area map. More often than not, because the site is found during exploration, this 'placement' will involve determining the direction of some point within the familiar area from the new site rather than *vice versa*. There are two main ways in which this can be done (Baker 1981)

The first method involves the use of information acquired during the outward, exploratory journey (i.e. *route-based navigation*) and several variations on this theme are possible. The animal could show *route-reversal* and retrace its outward journey exactly. Alternatively, it could 'cut corners' by moving along a succession of landmarks perceived and memorised, though not necessarily visited, during the outward journey. Route reversal is most simple when the outward journey is a straight line and oriented either:(1) along a *leading line* (e.g. a river, row of hills, hedgerow); (2) in a particular compass direction; and/or (3) with respect to some other suitable cue (e.g. gravity, light or humidity gradients, slope of ground). The chief disadvantage of route reversal (except when the outward journey is a straight line) is that it involves returning to the previous familiar area by what is often an uneconomical route. An alternative to route reversal that avoids this lack of economy is to monitor the twists and turns of the outward journey and to vector the direction of the pre-migration familiar area from the present position, either at intervals, or more or less continuously, during the tortuous outward journey. Upon discovery of a new site, therefore, the direction of the familiar area is already known and return can be direct, irrespective of how indirect the outward journey may have been. The twists and turns of the outward journey can be monitored in two ways, either by reference to external reference sources, such as a familiar landmarks and/or directional cues such as the Sun or the Earth's magnetic field, or by reference to some internal dead-reckoning or gyroscope system (Darwin 1873, Barlow 1964). Direction finding by the latter system is known as *inertial navigation* and is the mechanism incorporated into most modern jet planes.

The second method of determining the direction of the familiar area from a new site discovered during exploration is to make use of information available at the new site itself (i.e. *location-based navigation*) rather than en route to it. As the new site is outside the area ever-visited, use of the familiar area map to determine home direction can only make use of landmarks that can be perceived at a distance from both the new site and from within the previous familiar area. The mechanisms available depend on the nature of the landmarks and the senses by which they are perceived. Figures 10.6–8 indicate three possibilities, depending on whether the landmarks are visual/topographical features or air- or water-borne chemicals. All three models involve navigation based on relating the compass direction of familiar landmarks perceived at or from the new site relative to their compass direction from some site within the familiar area.

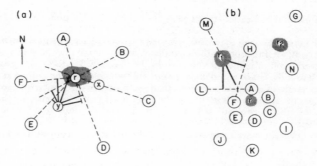

Fig. 10.6 One method of location-based navigation using familiar visual landmarks and a compass
A-F are gross visual landmarks that can be seen from a distance, such as mountains, hills, woodland, moorland, large rivers, lakes, etc. on land, or islands, coastlines, patches of different-coloured water, submerged topographic features, etc. at sea. r is a resource location. It is postulated that an animal, while at r, learns the compass directions of familiar landmarks. An animal that finds itself at y observes, among other things, a compass shift of A, F, D and C but a negligible shift of E and B. In order to return to r from y it is simply necessary for an animal to 'compute' which direction will restore all familiar landmarks to the compass direction that pertains at r. One way of doing this would be to compute the direction (solid lines drawn from y) that would restore by the shortest track each landmark to the compass direction appropriate to the location of r, and then move in the mean vector of these directions. Such a system can be used as long as at least one familiar landmark can still be seen and recognised, though accuracy of homeward orientation improves: (1) as the number of available landmarks increases; and (2) as these are spread around 360° of the visual horizon. A similar system would work for acoustic landmarks.
[Simplified from Baker (1978a). (This model contains elements of the independently developed 'mosaic map' postulated by Wallraff (1974) and the process of 'geodetic orientation' described by Merkel (1978)]

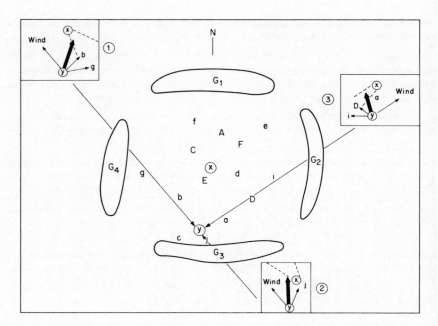

Fig. 10.7 A method of location–based navigation using familiar olfactory landmarks and a compass

It is postulated that while at x the animal learns to associate olfactory signature of winds with the different compass directions of those winds. Wind direction may be detected tactilely or by visual observation of cloud movements. In the diagram, G_1 to G_4 are gross olfactory features, such as hills, cities, woodlands, rivers, etc. which reach the animal at x on winds from a wider range of compass directions than the more localised olfactory sources, A–F and a–j. When the animal finds itself at y, the most potent olfactory cues are those of G_3, a and c which the animal associates with winds from the SSW while at x. Home direction is therefore NNE.

[From Baker (1978a), modified in accordance with the model first postulated by Papi *et al.* (1972)]

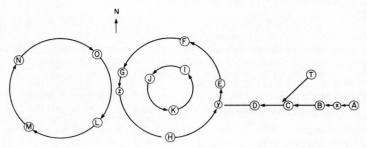

Fig. 10.8 A method of location-based navigation for an aquatic animal using familiar olfactory/gustatory landmarks and (optionally) a compass

Current direction is usually more predictable in aquatic systems than in terrestrial systems and it may not be necessary to include the use of a compass in the mechanism described below. Use of a compass, however, would undoubtedly increase the efficiency of the mechanism. A–O and T are stationary olfactory sources and x, y and z are locations

important to the animal as well as being olfactory sources. Consider an animal born at x that explores the length of the linear current and joins the gyral at y. As long as y can be re-located, x can also be relocated by swimming against the current for as long as any combination of D, C, T, B, x and A (but not A or T alone) contribute to the olfactory signature of the current in which the animal finds itself. If the signature of A or T alone is encountered, the animal should swim downcurrent until re-locating a signature that also contains D, C, B or x. Except that it is based on exploration and a spatial memory rather than sequential imprinting, and is thus more flexible, this part of the model shares many points with the model suggested by Harden Jones (1968). Having arrived at y from x, an explorer receives two pieces of information. First, y is on a gyral with a signature HEFG. Second, y is located at a specific point on that gyral where the contribution of each of these components has a characteristic relationship (in this case H probably makes the greatest contribution and E the least on the assumption that concentration decreases with distance from source). y may also be learned as that point on the HEFG gyral at which the current flows to the north. Suppose on its first exploratory migration from y the animal migrates at an angle anticlockwise of the downcurrent direction (or west) and encounters water with a JIK signature. Suppose that the animal continued through the JIK gyral. At z it would re-encounter the HEFG gyral. Sufficient information is immediately available for the animal to determine that if it remains with the gyral it will return to y, even though it has never visited z before. By learning the spatial relationships of different gyrals and by learning the particular contributions of different olfactory sources that identify a specific location (e.g. z is just downcurrent, or south, of the point where G makes its maximum contribution to the HEFG gyral and that from this point movement at an angle clockwise of downcurrent (or west) enters the animal into the MNOL gyral) the animal can readily establish a huge familiar area.
[From Baker (1978a)]

No serious suggestion has been made for most animals that they have the ability to use location-based navigation in order to determine the direction of a position within their familiar area following experimental release beyond the limits of their familiar area map. For birds, however, and to some extent for bats, such ability has been proposed. Indeed, it has been suggested that birds have a *grid-map* which can be used from any point on the Earth's surface. This map may be learned (Wiltschko and Wiltscho 1978) or may be innate and is considered to be based on some combination of astronomical or geophysical cues.

Whether an animal has arrived at an unknown site as a result of exploration or experimental displacement, the problems with which it is confronted are the same. In either case, the animal has to determine the direction of its home site by some form of navigational mechanism. There is no fundamental difference between the two. Displacement-release is essentially enforced exploration and it seems reasonable to expect the animal to show similar behaviour in both situations. We started this chapter by noting that in people's minds navigation had always been inextricably linked with long-distance migration. We can now see that it is exploration with which navigation shares such an intimate relationship.

The link with long-distance migration, as we shall see later, exists only insofar as this also involves exploration. Freed from doubts that displacement-release experiments are invalid and that experiments on Homing Pigeons tell us nothing of 'true' navigation, we can now take that elusive step forward; to see if this new view of animal navigation as part of exploration can start to unravel the whole mysterious behaviour.

Assistance in this step comes from an unexpected quarter, though it emerges naturally from linking navigation with exploration and life within a familiar area. Man lives within a familiar area built up by exploration. Why, then, should he not also navigate in a way comparable to other animals? Indeed, we can go further. At a crude approximation, the sizes of the familiar areas of Men and Homing Pigeons must be of similar orders. Moreover, in general terms, the sensory apparatus of the two species is similar, a point to which we shall return later. Pigeons, because they are able to rise up into the air, can move in straighter lines and perceive landmarks from a greater distance than Man, who has to seek suitable vantage points in order to obtain comparable views of the distant horizon. Apart from this, however, the navigational problems needing to be solved by individuals of the two species as they explore and build up their familiar area seem to be similar. We might therefore expect evolution to have produced similar navigational ability in the two species (see Baker 1981 for longer discussion).

Countless papers and books on bird navigation include somewhere or other a statement along the lines that: 'if humans were treated like Homing Pigeons and transported to an unknown destination, they would have no idea of the direction of home'. Yet to the best of my knowledge nobody ever tried this simple experiment to see if it was true. Over the past few years I have been carrying out such experiments, and in the next few paragraphs compare the results obtained for humans with those obtained for Homing Pigeons. The results are surprising and shed light on both species. We proceed by looking first for evidence of route-based navigation, then of location-based navigation, and finally by considering whether animals can navigate when displaced beyond the limits of their familiar area map.

The first evidence for route-based navigation came from work on amphibians (see review by Ferguson 1971). When displaced in open containers, upon release the animals oriented in the compass direction opposite to the direction of displacement. If only one leg of the displacement was in an open container, orientation upon release was in the compass direction opposite to the direction of displacement during that leg taking no account of the displacement direction during the leg that the Sun was not visible. Amphibians displaced in light-tight containers showed no homeward orientation upon release. In retrospect, as Salamanders at least have access to a geomagnetic compass (Phillips and Adler 1978) and in view of results for pigeons, it is surprising that even in light-tight containers, as long as they were not iron, there was no indication of route-based navigation.

The switch of interest from location-based to route-based navigation in the study of Homing Pigeons has been a recent event (Wiltschko and Wiltschko 1978, Walraff et al. 1980). It began during the first half of the 1970s with work in Italy which showed that pigeons picked up familiar smells during displacement and used this information to monitor their displacement route. This was shown by dog-leg displacements (Fig. 10.9), pigeons being inclined to bias their orientation upon release in a direction opposite to that of the first leg of a dog-leg displacement, but only as long as they could smell the air through which they were being displaced. A number of elegant experiments (see recent papers by Papi *et al*. 1978 and Baldaccini *et al*. 1978) involving manipulation of the direction from which smells reached birds at the home site have also been carried out.

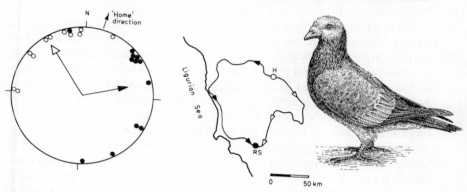

Fig. 10.9 The dog-leg or detour technique for demonstrating route-based navigation for animals that take most notice of the early phase of displacement: Homing Pigeons in Italy H, loft; RS, release site; line with open and solid arrows, two different routes between H and RS. In the left-hand diagram, open dots show the disappearance direction upon release of birds transported by the open triangle route; solid dots by the solid triangle route. Mean angles for disappearance directions (open and solid headed arrows) reflect the first leg of the displacement journey.
[Re-drawn, with simplification, from Papi *et al*. (1978)]

It now seems clear that, in some areas at least, birds at their home site note the direction from which they receive particular smells and can use this information during displacement to monitor displacement direction. In other areas olfactory cues seem sometimes to be used and sometimes not (Hartwick *et al*. 1978, Schmidt-Koening and Phillips 1978), perhaps depending on how different are the smells arriving from different directions at the home site.

Often birds displaced in light-tight aluminium containers and without access to olfactory cues from the environment can also monitor outward direction. The ability to monitor direction disappears, however, if the birds are displaced in an iron box or in a box in which the Earth's magnetic field has been disrupted (Kiepenheuer 1978, Wiltschko *et al*. 1978). It is also worth contemplating that when Mallard, *Anas platyrhynchos* are displaced from Slimbridge the first leg of the journey is almost always towards the south-east. Is this perhaps a factor in the 'nonsense' orientation to the north-west shown by these birds?

The evidence obtained over the past few years is now strong. Birds can monitor displacement direction. Although they may give greatest weight to the first leg of their displacement, they show some ability to vector the final direction even when the outward journey is not straight. Presumably visual cues are used when available. When they are not available, some combination of olfactory and geomagnetic compass cues is used, depending perhaps on experience at the home site and availability during the outward journey. Clear evidence for route-based navigation using a geomagnetic compass has also been obtained for the European Woodmouse, *Apodemus sylvaticus* (Mather and Baker 1981).

What of humans: can they employ route-based navigation? Using students from Manchester University, I have been displacing groups of between 5 and 11 individuals to 'release' points between 6 and 52 km from the University (Baker 1981). In order to test for route-based navigation, each individual was asked upon arrival at the release site to estimate the direction of the University (by describing it as a compass direction). The van in which displacement occurred had glass windows and not surprisingly, individuals who had been able to see through these windows during the outward journey were able to estimate the direction of the University with a high degree of accuracy. Most experiments, however, were carried out using students that were blindfold during the entire outward journey. These journeys were fairly tortuous and usually involved a dog-leg. The early results are shown in Fig. 10.10a and suggest a facility for route-based navigation even in the absence of visual cues during the outward journey. At the suggestion, and under the supervision, of Yorkshire Television, a one-off experiment was carried out at Barnard Castle, County Durham. Location, subjects, and equipment were all suggested and provided by Yorkshire Television and were unknown to the author. The results of this experiment are shown in Fig. 10.10b and suggest further evidence for route-based navigation in the absence of visual cues.

In these experiments, the subjects themselves could offer no description of the mechanism they used to make their estimate. Most claimed to have tried to follow the route on a memorised map of the area but found they had to give up after a few kilometres. On sunny days, a few claimed to have tried to use the heat of the Sun shining through the windows of the van or coach. Most, however, felt that they did not know the direction and were surprised when their 'guess' turned out to be so close to the home direction. In fact, there is evidence that neither memorised-map following nor detection of the Sun's direction was the primary mechanism of route-based navigation in these experiments. The dog-leg displacement prevents short-distance memorised-map following from being effective. Moreover, there was no deterioration in accuracy of homeward orientation with increasing distance of release; nor were individuals who, upon removing their blindfolds, found they knew their location, more accurate than those that

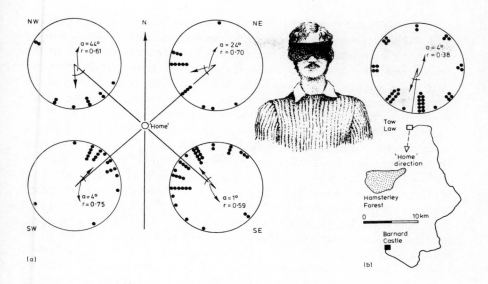

Fig. 10.10 Route-based navigation by humans blindfold during the outward journey
(a) Manchester series. 'Home' was Manchester University. Release points were between 6
and 52 km from the University, though outward journeys were longer. Data are lumped
for releases in four different compass quadrants relative to the University. In all, 64
individuals have been tested, spread over 8 different release sites on 11 different occasions.
Each dot is the verbally stated estimate (N, SE, WSW, etc) of home direction from the
release point by an individual (expressed in the diagram relative to the home direction).
Arrows show mean vectors.
(b) One-off Barnard Castle experiment. Solid square, 'home'; open square, release point;
wavy line, displacement route. Each dot is an individual's estimate of home direction, made
as before while still blindfold, but in this case by writing on a card rather than verbally.
[Compiled from Baker (1981)]

did not. Finally, displacements that took place under completely overcast
skies produced homeward orientation as accurate as displacements under
more or less continuous sunshine (Baker 1981). The mechanism of route-
based navigation by humans, therefore, was elusive. The break-through
came on the second of Barnard Castle experiments (Baker 1980c). In this
experiment, the subjects were divided into two groups. One group placed
bar magnets at the back of their heads, in the elastic of their blindfolds. The
other, control, group placed dummy brass bars in the same position. All of
the subjects thought that they were wearing magnets. The results obtained
are shown in Fig. 10.11. After the first leg of the dog-leg, these individuals
wearing magnets were significantly anti-clockwise of the home direction
by 74°, whereas the controls had a significant vector in the home
direction. After a sharp turn and further displacement along the second leg
of the dog-leg, the group wearing magnets were disoriented, producing
estimates of home direction not significantly non-uniform. The control
group again produced an accurate estimate of the home direction.

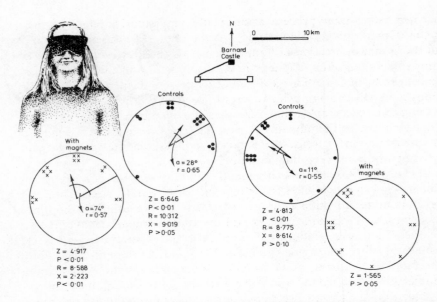

Fig. 10.11 The possible use of a geomagnetic compass in route-based navigation by humans blindfold during the outward journey
A group of 31 humans was subjected to a dog-leg displacement from Barnard Castle (■) as shown. At two points (□) on the dog-leg they were asked, while still blindfold, to write down an estimate of home direction. The group consisted of 15 individuals who wore bar magnets on their heads and 16 control individuals who wore brass bars. All subjects thought they were wearing a real magnet. Each dot or cross is the estimate of home direction made by one individual. Line from centre of circle shows true home direction, arrow shows mean angle for the estimates made if this was significantly non-uniform. z and its P-value tests for uniformity. R, X and their P-value test whether the mean angle is significantly different from the home direction. The experimental group wearing magnets show no indication of being able to monitor displacement direction accurately.
[Re-drawn from Baker (1981)]

Since the Barnard Castle experiment, my Manchester team has carried out numerous further experiments into the new-found human magnetic sense. The early part of the story was written up as a book (Baker 1981) which concentrated on the involvement of the human magnetic sense of direction in route-based navigation. More recently, we have obtained evidence that the human magnetic compass: (1) is reset each night in a way that is influenced by the direction in which we sleep; and (2) is less efficient at some times of day than others (Baker and Mather MS). We have also obtained evidence that the compass can be disrupted by quite small changes in magnetic intensity. To understand the significance of this, however, we have to enter the realms of magnetic grid maps and magnetic storms. First, however, we have to appreciate the difficulties of experimental design that result from the possibility that animals could, *and often do*, use route-based navigation.

The favoured interpretation of the navigation of pigeons at present, is that it is a two-stage process, as originally suggested by Kramer in the 1950's (Gould 1980, Walcott 1980). At the release site the pigeon works out the compass direction of home Stage 1). It then refers to some set of compass cues (e.g. sun, magnetic field) and translates its Stage 1 compass estimate into a direction for travel. This translation is Stage 2 and is a simple compass response. Stage 1, however, can be achieved in two ways, either by reference to a 'map' (either a familiar area map or a grid map) or as a result of following the outward journey. In other words it can be achieved by either location- or route-based navigation. We can illustrate this again by comparison between humans and pigeons.

During the Manchester experiments, the human subjects were asked, after making their verbal estimate of home direction, to point while still blindfold in the direction of home. The scatter of the estimates about the mean angle was much greater than for the verbal estimates. Nevertheless, the orientation was significantly non-uniform and the mean angle was not significantly different from the home direction. Even while blindfold, therefore, the humans could have started walking with a statistically significant bias toward the true home direction, even though many would have set off in the wrong direction. The way the subjects did this was to use their Stage 1 compass estimate of direction from home and then to relate this 'mental' direction to a proposed direction for travel by trying to use some external cue (Stage 2). This involved making reference to wind direction (which many had noted at the home site before displacement began and before being blindfolded and which was then detected at the release site by the feel of the wind on the face) and/or the direction of the Sun (from the feel of the Sun's heat on the face), and/or the magnetic field.

An essentially similar experiment has been carried out on Homing pigeons. Instead of being released while blindfold, however, pigeons were released while wearing frosted-glass contact lenses (Schmidt-Koenig and Schlichte 1972, Schmidt-Koenig and Keeton 1977). Such pigeons surprised everybody at the time by being able to set off after release with a significant bias toward the homeward direction. Indeed, many returned near to, and some even entered, their home loft. I remain unconvinced that these pigeons could not see the outline of horizon landmarks (Baker 1978a), but if we accept as the authors suggest that they could not, the experiment is very similar to the experiment on blindfold humans. Suppose the pigeons had employed route-based navigation for Stage 1. They could return to the vicinity of their loft as long as they had access to a compass at the release site. They could have used the geomagnetic compass, but clock-shift experiments have shown that in fact they used a Sun compass, the position of the Sun being visible through the contact lenses.

Students of navigation have spent decades trying to study the location-based mechanisms involved in Stage 1, never realising the full significance of the possibility of route-based navigation. This was a pity. Unfortunate though it is, it means that nearly all navigation research to date tells us little about Stage 1 of navigation. Students of pigeons tried to control for the possibility of inertial navigation during the outward journey (see review by Matthews 1968) by anaesthetising the birds, placing them on turntables during the outward journey, or damaging or plugging the semicircular canals. The last technique does not control for inertial navigation by movement of the viscera (Delius and Emmerton 1978); the turntable experiments do not control for route-based navigation by olfactory or perhaps even geomagnetic cues; and anaesthetisation produces erratic results. In any case, all such birds were released under conditions of good visibility and sunshine at reasonably short distances from home at sites where location-based navigation should have been straightforward. In other words, such experiments were unlikely to lead to reduced homeward orientation, whether or not inertial navigation was important in route-based navigation. It is often pointed out that pigeons sleep during displacement. So, too, did some of my humans (Baker 1981). Yet both groups can still orient towards home at the release site.

It seems a drastic step to say that we cannot use any of the navigational data built up over the years in our study of Stage 1 if the experimenters did not control for route-based navigation, yet we can do nothing else. The reason can be seen most clearly if we consider the final part of the Manchester experiment on humans. After pointing while still blindfold, each individual was asked to remove the blindfold, to look around, and then to point once more in the direction of home. Release sites were chosen for an open, usually rural, view so that in many directions the visual horizon was at 30 km or so. There were no signposts, buses or main roads visible and the tall buildings of Manchester City Centre were always hidden from view by either the horizon or by strategically-placed trees or houses, etc. This was a test of visual location-based navigation; but how could it be studied, given that when each student left the van they already had in their minds an estimate of the direction of home often derived from route-based navigation? After making a visual estimate of home direction each person was asked what cues he or she had used in making that estimate. Half claimed to have used, among other things, their route-based, verbally-expressed, estimate of their direction, and the correlation between their two estimates was so good that this claim could be believed. The other half used as a starting point, not their verbal estimate, but the direction in which they pointed while still blindfold. Again the data supported this. All individuals, therefore, began their attempt at visual location-based navigation with a preconceived idea of home direction based on route-based navigation. It was found that, as long as they had access to a compass, either the Sun or the wind (for those that had noted wind direction before displacement began), the subjects could produce a visual estimate as accurate as their verbally stated route-based estimate. Under calm, overcast conditions the visual estimate was much worse than the verbal estimate (Baker 1981). Yet this tells us nothing about visual location-based navigation; only how accurately the subjects can use naturally occurring compasses to translate their route-based, estimate into a proposed direction for travel once they can see. In order to study visual location-based navigation, we have to look for an improvement in the visual estimate over the earlier, perhaps route-based, estimate. Yet this earlier estimate is usually so good that there is little room for improvement. In fact, in the human experiments, because the blindfolded Stage 1 estimate was known and could be distinguished from the visual estimate, any improvement of the latter over the former could be measured. It was found that under conditions of sunshine and/or detectable wind plus visibility good enough to see distant (about 30 km) horizon features, the visual estimate did show an improvement over the verbal estimate, at least in males. There is, then, an indication of location-based navigation by humans based on distant landmarks (mainly hills or their absence) and a compass, though again the subjects themselves could give no clear description of the way they used these features to make their estimate.

Visual location-based navigation could only be demonstrated for humans because blindfolded and sighted estimates could both be obtained and compared. A similar study has succesfully been carried out for mice (Mather and Baker 1981). In the absence of both of these pieces of information it is impossible to study visual location-based navigation. If the non-visual Stage 1 estimate cannot be obtained, the only way to study visual location-based navigation, is to design experiments in such a way that possibility of route-based navigation in Stage 1, whether by inertial navigation or reference to geomagnetic, olfactory or other cues, is totally disrupted (Mather and Baker 1981, Wallraff 1980). The worst that can be done is simply to displace an animal and release it. Whether it homes or not, the result gives no indication of the involvement of either route-based or location-based navigation in Stage 1. As 99.9 per cent of all navigation experiments in the past have fallen into this category, we now find that there is a negligible amount of critical data available concerning Stage 1. However, it is easy to be wise in retrospect; this could not have been foreseen by the experimenters concerned.

When the route-based navigation system of pigeons is disrupted by displacing them in iron boxes or manipulating the magnetic field and/or by altering the direction from which familiar smells reach birds at the home site; such birds show reduced homeward orientation when released but show little reduction in homing success. Evidently, therefore, pigeons do show location-based navigation but this has not yet been studied; indeed it will be difficult to do so if the study has to rely on analysis of homing success. Pigeons released at the site of magnetic anomalies are disoriented, and this could suggest the involvement of magnetic cues in location-based navigation at Stage 1 (Walcott 1980). So, too, could the influence of magnetic storms on pigeon orientation, (Keeton, Larkin and Windsor 1974). Pigeons that have already set off toward home fly straight over magnetic anomalies without deviation (Walcott 1978). An ingenious technique whereby a small camera is placed on a pigeon's head so that it records (roughly) the sites scanned by the pigeon's eyes (Köhler 1978) suggests that after release they systematically scan the horizon and take special note of horizon features. Scanning is more intense at visibility of only 12 km than at 30 km. This implies visual location-based navigation is also involved in Stage 1 for pigeons.

Navigation by a rodent has only just been demonstrated (Fig. 10.12) yet already it has been possible to separate to some extent the roles of route-based and location-based mechanisms in Stage 1 (Mather and Baker 1981). Previous studies on small mammals, however, employed only homing (see review by Bovet 1978), which is just as susceptible to confusion between route-based and location-based mechanisms. Woodmice, *Apodemus* spp show 59 per cent homing from 250 m and 17 per cent homing from 750 m. Such animals could have homed simply by reversing the direction they obtain by route-based navigation, having had no idea of their actual location upon release. The decrease in homing success with distance could reflect any number of factors, not least the effects of predation during the attempt to return.

Almost all of the navigation studies on bats have also studied only homing, the major exception being that by Williams and Williams (1970), who used radio-telemetry to monitor the behaviour of *Phyllostomus hastatus* after release (Fig. 10.13). Bats have been subjected to all combinations of blinding, blindfolding and earplugging in attempts to determine the relative importance of visual and echolocation senses in navigation. The involvement of both senses has been indicated (though whether in route-based or location-based navigation is unknown) and vision seems to become more important with increasing release distance. This conclusion has been criticised by Mueller (1968), who argues strongly that the results could equally be due to differential predation on blinded and ear-plugged bats and that blinded bats are at disproportionately greater risk to daytime predators when the bats are released at greater distances. Some experimenters have released bats inside and outside their presumed familiar area. These results have been contradictory, but as the possibility of route-based navigation was not controlled, navigation from outside the presumed familiar area (e.g. Leffler, Leffler and Hall 1979) tells us nothing about whether familiar landmarks are involved in location-based navigation. The suggestion from Fig. 10.13, though again route-based navigation was not controlled, is that location-based navigation only occurs if familiar landmarks are visible on the distant horizon, a conclusion reminiscent of that found for human location-based navigation.

Navigation experiments on amphibians have so far controlled for route-based navigation only with respect to visual cues on the outward journey. As some, at least, have access to a geomagnetic compass (Phillips and Adler 1978) we might expect amphibians, like mice and humans, to use this compass in route-based navigation. Critical data are not yet available. A number of experiments on a variety of toads and salamanders involving plugging noses and blinding have shown that both vision and smell are

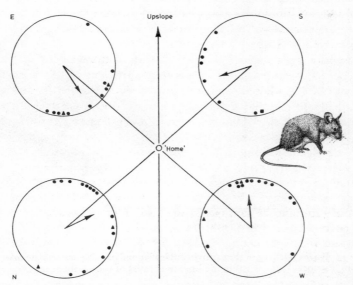

Fig. 10.12 A demonstration of navigation by rodents subjected to experimental displacement
'Home' was the Longworth trap in which the individual had been captured a few hours before displacement and testing in an orientation cage. Experiments were carried out in Gloucestershire, England, in summer and were on a steep slope running from north-west to south-east. Data are lumped for displacements in four different compass quadrants relative to the trap in which the individual was captured. In all 19 individuals were used, (●, Wood Mouse, *Apodemus sylvaticus;* ■, Yellow-Necked Wood Mouse, *A. flavicollis;* ▲, Bank Vole, *Clethrionomys glareolus*). Each dot indicates the mean angle of orientation for an individual in the orientation cage expressed relative to the home direction. Only data for release distances between 6 and 80 m from 'home' are shown. Beyond 80 m (nearly 3-times the radius of the home range), homeward orientation disappeared.
[Re-drawn from Mather and Baker (1981)]

involved in navigation following displacement-release, though whether route-based or location-based or both is unknown. Some anurans can recognise the water from their spawning site by its chemical signature (Grubb 1973). Perhaps the most informative experiments, however, have been the least conventional (see review by Ferguson 1971). Frogs and toads caged on an unfamiliar shore show evidence of being able to orient to their new surroundings within 2 h and the process is nearly complete in 24–48 h. The role of a compass in this process was shown for Southern Cricket Frogs, *Acris gryllus*, which failed to orient if not allowed to see the daytime sky. Once allowed to see the daytime sky, however, the frogs could orient to the new shore in both day and night tests. Chorus Frogs, *Pseudacris triseriata*, caged for several days near a breeding chorus of conspecifics, then

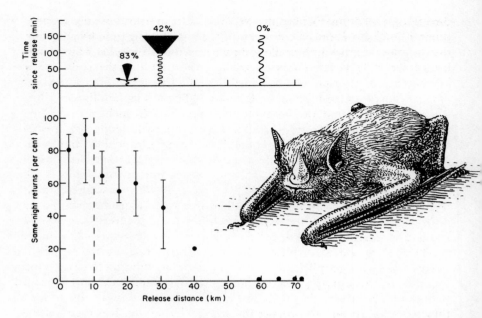

Fig. 10.13 Radio tracking of the carnivorous leaf-nosed bat, *Phyllostomus hastatus*, released at varying distances from roosts in the limestone caves of the northern mountain range of Trinidad

Bottom diagram shows variation in homing success with release distance (mean and range over different experiments at each distance). Vertical dashed line shows the limit of the normal foraging area, but not necessarily the familiar area, as determined by radio tracking. Top diagram shows the results of radio–monitored homing experiments. Wavy line indicates period of 'random' movement after release. Black segment shows the angular scatter of homing orientations and the number shows the percentage of bats that had returned to their home roost after the time indicated. The visual acuity of the species is such that the mountain range in which the home roosts were located seems likely to be visible from 10 to 20 km, visible with difficulty from 30 km, and not visible at all from 60 km. [From Baker (1978a), after Williams and Williams]

moved and released in a different arena, selected a compass course that would have taken them to the chorus from the original site, but not from the site of testing. These experiments seem to show that the familiar area map of amphibians consists of stored information concerning landmarks, whether visual, olfactory or acoustic, integrated with appropriate compass information.

In the past, herpetologists have noted that many amphibians home in variable winds or winds that are not blowing from the home site to the animal. They considered this to show that familiar olfactory landmarks were not involved in navigation. The model produced by Papi *et al.* (1972) for pigeons shows this argument to be invalid (Fig. 10.7).

Displacement—release experiments have to be carried out with some precision in order to give information on location-based navigation. If an animal is released within its familiar area it already knows the compass direction or other instructions that will take it from the release site to some other 'home' site. It can, therefore, home without navigation by adopting the compass direction or by following the leading line or other route that it has learned will take it from the release site to home. Given that route-based navigation can in some way be controlled, the true test of location-based navigation comes from releasing the animal beyond its familiar area, yet still within the boundaries of its familiar area map. In such situations the animal should show location-based navigation perhaps along the lines of the models presented in Figs. 10.6 – 8. These are the models to be tested. What little information can be gleaned concerning location-based navigation from the data currently available is quite consistent with these models without actually proving them to be correct. If we release the animals beyond the limits of even their familiar area map, then they should be disoriented, unable to locate home direction, as long, that is, as route-based navigation has not occurred and they do not have access to a grid map.

The idea of a grid map has only seriously been postulated for bird navigation. Some geophysical or celestial factors are thought to be detected by birds, in effect giving a bicoordinate grid over the Earth's surface such that no two points are the same. The bird learns or is innately able to read these coordinates and memorises those of its home site. It is then able to compare coordinates at any position on the Earth's surface with those at the home site and thus determine home direction. On the grid-map system, the bird is able to navigate from any release position, no matter how far it is beyond the limits of its familiar area map.

The concept of a grid-map is simple and appealing and for two decades it seemed that the only problem ornithologists had to solve was the identification of the cues by which the coordinates could be read. The most famous suggestion was the Sun-arc grid map theory (Matthews 1955, 1968). It was argued that, like humans using a clock and sextant, birds could determine their latitude from the Sun's altitude and their longitude from the time of day at the release point. By comparing these two measures at the release point with those as they should be at the home site, the bird could, using relatively simple rules, determine the direction of home. Local time could be detected by noting how far along its arc the Sun had travelled or by noting its rate of climb or descent. This could then be compared with time at the home site as indicated by the bird's internal biological clock. This clock, once phased to time at home, cannot be rephased (or so it is postulated).

A variety of authors have attempted to disprove the Sun-arc model by the simple device of shifting a bird's clock by a small amount (Fig. 10.5). According to the Sun arc model, the bird should interpret such a shift in its clock as a shift in longitude. Suppose the bird's clock is shifted in such a way that it thinks it has been displaced 200 km to the east of the home site. If the bird is then displaced 100 km to the west, beyond the supposed limits of its familiar area map, it should, according to the Sun arc model, still fly to the west. In all experiments so far, the birds home more or less normally with just a slight shift in disappearance direction consistent with the compass shift that should also accompany such a clock shift. However, such experiments have not controlled for route-based navigation and as a result cannot be considered to be an unequivocal test of anything. I have interpreted the results of clock-shift experiments as supporting the model presented in Fig. 10.6, but again the failure to control for route-based navigation devalues this support.

Matthew's Sun-arc model was not the only grid map suggested. If birds could detect two components of the Earth's magnetic field they would again have a grid map. The best of such maps would perhaps be obtained by monitoring some, or all, of *inclination* (the angle which a magnetic compass makes with the horizontal), *declination* (the angle between magnetic and geographic north), the *total field intensity*, or the *intensity of the vertical component* (Gould 1980). Birds are known to detect the angle of inclination, at least in a gross sense, and should have the ability to detect declination. The fact that pigeon goal orientation is influenced by magnetic storms has been interpreted as indicating that the detection of intensity is also involved in Stage 1 (Keeton *et al.* 1974, Larkin and Keeton 1975), on the assumption that changes in intensity and declination due to magnetic storms are too small to influence a magnetic compass. The most exciting and latest discovery by the Manchester team, is that goal orientation by people is disrupted by magnetic storms (Baker and Mather MS). For both species the influence of storms is suppressed if there is an artificial magnetic field through the head during the experiment. This suggests that the effect is magnetic and not the spurious result of, for example, meteorological changes that might also correlate with magnetic storms. However, since it now seems that the magnetic compass is also influenced by magnetic storms, the observed influence of magnetic storms and anomalies on goal orientation can no longer be considered as evidence for a magnetic grid map based in part on intensity (Baker and Mather MS).

There is similarly meagre support available for other postulated grid maps. In an important review paper, Whiten (1978) presented evidence that, albeit under unnatural operant conditions, pigeons may indeed use the Sun's altitude as a navigational clue. There was no evidence, however, that they could use the Sun's arc to monitor longitudinal displacements. Without this second coordinate a useful grid map does not exist. Night-sky grid maps, based on the arc of the Moon or of individual stars, can also be postulated but as yet there is no evidence that the necessary coordinates can be detected (but see the discussion of the use of zenith stars by humans; Baker 1981)

Evidence that birds can navigate by using a grid-map when released beyond their *familiar area* map is not easy to obtain and I have suggested

(Baker 1978a) that most authors underestimate the size of the familiar area map, often confusing it with the much smaller familiar area. This is certainly true if the limits are set by infra-sound landmarks as suggested by Kreithen (1978). Most releases of pigeons and other birds may have been within the birds' familiar area map. Releases that certainly were not, such as trans-atlantic displacements between Germany and the United States failed to control for route-based navigation. We may wonder in what direction the birds would have flown upon release had they been displaced from Germany to the United States along an eastward route instead of the westward route actually taken.

Although the possibility that birds might use a grid map is exciting more interest now than at any time in the past, there are experiments that seem to argue against any form of grid map. The most enigmatic of these was performed in Germany. Kramer (1957,1959) and Wallraff (1966) confined pigeons in an aviary (3 m high and 10 × 6 m at the base)from birth and prevented them from seeing landscape features, either by keeping the aviary in a bomb crater or by giving the aviary opaque sides. When released 150 km to the south of the home loft, no birds returned, but even more importantly 68 recoveries elsewhere gave no hint of homeward orientation (Fig. 10.14). Other birds, similarly caged but given a view of the distant horizon, gave a 7.2 per cent return (n = 125) with other recoveries giving a homeward bias. Birds with a single gap in the opaque walls permitting a total view of 140° of the horizon showed zero returns (n = 165). Others with five gaps, each 1/15th of the perimeter, so that the complete horizon was visible, but not simultaneously from any single position, gave a 5.8 per cent return. Allowing the birds to see the horizon but not the Sun gave ambiguous results.

These early experiments seemed to show that in order to be able to navigate, pigeons have to be able to see distant landmarks from their home site around 360° of the horizon. Navigation is further improved if these landmarks can also be related to the Sun, probably for use as a compass. It seemed, then, from these experiments, that a familiar area map involving in this case visual landmarks and a Sun compass is necessary for navigation to occur. Previously (Baker 1978a, Fig.33.20) I interpreted these data on the basis of location-based navigation and took the results to support the model shown in Fig.10.6. This support is no longer valid, partly because the experiments failed to control for route-based navigation. Mainly, however, it is not valid because of the most recent experiment by Wallraff (1979). He took the simple step of raising his birds in an aviary with glass walls. Such birds could not navigate upon release. Whatever the walls were screening out in the earlier experiments, therefore, it was not horizon landmarks, but something that cannot be perceived accurately through glass. Polarised light is a possibility.

This experiment also poses the quesion, if pigeons can use a geomagnetic

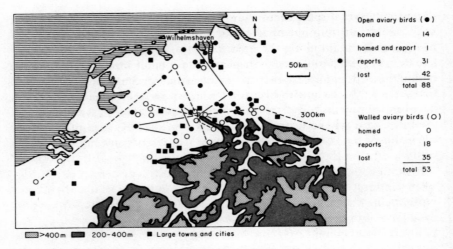

Fig. 10.14 The effect of reduced pre-displacement experience on the navigation of Homing Pigeons

Pigeons at a loft just south of Wilhelmshaven were in effect divided into two groups. Both were retained in an aviary from fledging onwards. One group was maintained in an open aviary and had visual access both to the Sun and to the horizon. The other group was maintained in a walled aviary such that although the Sun was visible the horizon was not. Both groups were then released (separately) at R with the results illustrated. Lines (solid, open aviary birds; broken, walled aviary birds) are shown only when a bird was reported more than once.

[From Baker (1978a), after Kramer and Fullard]

grid map, why could they not home even after being raised in a walled aviary? Wiltschko and Wiltschko (1978) have suggested that birds have to build up a grid-map in much the same way as they build up a familiar area map. As they fly around, ever further from home, they could detect gradients (e.g. of magnetic field intensity which in the northern hemisphere increases to the north). Having detected these gradients, it should theoretically be possible to extrapolate from them and thus recognise locations further from home than the bird has ever visited. This does not explain, of course, why birds raised in an aviary *can* home as long as the aviary is not constructed with opaque walls or glass. It does, however, explain away why they may have been unable to use a geomagnetic grid-map.

As far as location-based navigation 'within' the familiar area map is concerned, however, experiments on caged Starlings, *Sturnus vulgaris* have shown a need to see both distant landmarks and the Sun in order to recognise location (Cavé *et al.*1974). As these experiments effectively controlled for route-based navigation by showing that birds were disoriented after displacement if distant landmarks and the Sun were not both visible they do provide some support for visual location-based navigation, perhaps along the lines shown in Fig. 10.6.

The final conclusion, then, is that animals probably do use a combination of route-based and location-based mechanisms in Stage 1 of navigation (Stage 1 = determining the Compass direction of home). Visual location-based navigation requires reference to a familiar area map consisting of familiar landmarks and compass information. The only grid map beyond the limits of the familiar area map for which there is any support among ornithologists is a geomagnetic grid map. It is still possible that all navigation is based on a familiar area map, either as the initial, oriented mental springboard for route-based navigation or as the means for location-based navigation. Suppose this were true, however, what are the dangers of an animal becoming lost?

The great advantage to an animal of using a world-wide grid map is that theoretically it can never become lost, as long as it can perceive its coordinates. Obviously, if all cues disappear as a result of thick fog, magnetic storms, or sensory impairment, for example, any animal can become lost whatever navigational system it normally uses. On the least-navigation system, some cues are used in preference to others because they are more efficient and/or easier to use, in short, because they require the least navigation. As the preferred cues disappear or are removed experimentally, navigation should become more difficult, though not necessarily less accurate, until eventually there are simply not enough cues available for an animal to find its way around. Explorers using location-based navigation, because they require distant cues, are likely to become lost first; route-based navigators requiring compass information are likely to become lost next; and finally even individuals within their familiar area, who could resort to detailed short-distance information if necessary (Chapter 5), could still become lost if conditions were to deteriorate sufficiently.

All this is obvious . Our main interest, however, is in whether there is a danger of animals becoming lost during exploration even if conditions do not deteriorate, simply because they are using only route-based information and a familiar area map. The answer should be that, with few qualifications, they do not get lost.

Imagine an animal that leaves its familiar area on an exploratory migration. As long as it remains within its familiar area map, of course, there is no difficulty in returning to appropriate parts of the familiar area and incorporating a new site within that area. All this can be achieved by location-based navigation (Figs. 10.6–8). Remember, too, that the familiar area map can extend a long way from the familiar area, depending on the senses used to perceive the direction of distant landmarks. We are interested, though, in the situation in which an animal explores beyond the limits of its familiar area map, as well as beyond the familiar area. For many animals this may never happen, because as they explore they will also extend their familiar area map. Indeed, it is not so easy to envisage situations in which animals would not extend their familier area map in this way and be forced to fall back on route-based navigation. However, animals such as amphibians, rodents and terrestrial reptiles that have to explore through dense vegetation, or even larger animals exploring, say, through forest, may have to travel some distance without being able to perceive distant landmarks, visual or olfactory. If such explorations carry the animal, perhaps inadvertently, beyond its familiar area map, route-based navigation would be an efficient means of relocating familiar landmarks.

Perhaps the most likely function for route-based navigation using non-visual cues in the real world is to counter situations in which the familiar area map effectively shrinks in size while exploration is in progress. For example, animals using distant visual features may encounter fog during exploration. Apart from this, the obvious situation for birds in which route-based navigation becomes invaluable is when an individual en-counters winds contrary to the direction in which it wants to explore and too strong for it to maintain its required track. Normally, such a bird would settle, but if it were a land-bird over water or a sea-bird over land it would probably do better to allow itself to wind-drift until it could settle (Williamson 1955). Examples of wind-drift are probably less common than previously supposed because often apparent examples simply reflect birds taking advantage of air currents to explore a direction economically. However, enforced wind drift undoubtedly does occur, though wind-drifted birds, as well as downwind explorers, might be expected to extend their familiar area map as they go. Even so, given other unfavourable conditions, such as low cloud or mist, so that visual landmarks are not available, route-based navigation during wind-drift may often be the only mechanism available.

Suppose an animal does travel beyond the limits of its familiar area map in circumstances such that it is forced to employ route-based navigation. Some time later it has to attempt to return to its familiar area. What are the chances that it will miss its familiar area map altogether and thus be 'lost'? Obviously this depends on how accurately the animal can employ route-based navigation and how large is its familiar area map. The relationship is a simple one to work out. If the distance of the animal from the nearest point at which it can perceive familiar landmarks is twice the diameter of its familiar area map, the animal has to be able to use route-based navigation to within $\pm 11.5°$; if ten times, to within $\pm 2.7°$; if 100 times, to within $0.29°$; if 1000 times, to within $0.03°$. Obviously, the danger of getting lost is greatest while the familiar area map is very small, such as when young. There is a clear advantage in building up the familiar area map gradually at first but to a reasonably large size before any long-distance explorations are attempted. This is exactly what most animals are found to do, whether humans, otters, rabbits or birds. The calculations suggest that explorations further than twice the diameter of the familiar area map are likely to be dangerous if the animal has to rely on route-based navigation, rather than extension of the familiar area map itself.

What does this mean in terms of real animals? We have already seen that, even without moving, an animal can have a relatively large familiar area map. A bird that does no more than fly up into the air and down again on a day of good visibility can have a familiar area map 100 km across. This means that with no further action the bird could probably be taken 250 km away from its starting point and still be able to home, first by route-based then by location-based navigation once it encounters familiar landmarks. If we accepted route-based navigation to an accuracy of $\pm$ 3° instead of 11.5°, then the bird could be taken 1000 km away and still be able to home. All this without ever having moved from its place of birth save to obtain a clear view of its distant horizon. If the bird then begins to explore around its starting point and develop an even larger familiar area map, it could go still further. We can take it, I think, that the combination of route-based navigation and familiar area map is likely to mean that few animals ever get permanently lost as long as they wait for conditions in which their familiar landmarks can be perceived from a reasonable distance and as long as they have access to a reliable compass for both route-based and Stage 2 of navigation.

We are now in a position to start to try and make sense of the relationship between navigation and long-distance seasonal migrations, beginning with birds. For this, I shall avoid the involvement of a geomagnetic grid map which still seeks convincing support. The challenge is to construct bird migration circuits on the basis of the exploration/familiar area model, if bird navigation consists only of route-based and location-based mechanisms, the latter being based only on a familiar area map. Wiltschko and Wiltschko (1978) attempted a parallel exercise but assumed that a familiar area map was used primarily around the birth site.

A young bird builds up a basic familiar area around its birth place by post-fledging exploration. Not only does it explore for potential breeding sites, it also builds up a familiar area map (based on visual and olfactory landmarks?). Even without moving from its nest site its familiar area map could be 100 km across (larger given an infra-sound map). It can explore within this area on days of good visibility without even noting where it is going. The route can be as tortuous as the bird likes. Upon discovery of a resource location, the bird can work out the direction of other resource locations by location-based navigation (Fig. 10.6). Thereafter, it can move between any two locations simply by orientation; if A is north of B then when leaving A fly due south. As the bird explores within this original 100 km area, the boundaries of its familiar area map are pushed outward because from the perimeter of its first area it can perceive still more distant areas. As long as a bird can extend its familiar area map as it explores it need never run the risk of becoming lost. Route-based navigation without landmarks becomes most important when visibility decreases. If this occurs rapidly, an individual can be exploring within its familiar area map one moment but be unable to perceive familiar landmarks the next. Suppose that such route-based navigation and the ability to reverse the direction thus determined continue to produce an accuracy of $\pm 10°$. On this assumption, the bird should endeavour never to be further away than twice the effective diameter of its familiar area map under current conditions of visibility, compass availability, etc. If this ever does happen, the bird should settle and wait for conditions to improve before attempting either to return or to continue exploration.

As post-fledging migration continues and the familiar area map grows larger and larger, the distance that a bird can explore beyond the limits of that map grows exponentially. To judge from ringing returns of young birds carrying out post-fledging migration, by the time of the autumn migration the young bird can have a familiar area map hundreds of kilometres across. Suppose the standard migration direction for the species is south, and that eventually the bird flies in this direction. Suppose again that it is able, if it so wishes, to reverse its outward direction to within an accuracy of $\pm 10°$. It could probably do better. Given this, however, it means that theoretically the bird could fly a first leg of 2000 km or so, taking no notice of its surroundings, and still be able to return to its familiar area. In fact, we should suppose that the bird extends its familiar area map as it goes; that either side of its migration route, distant landmarks are perceived and entered onto the familiar area map within the spatial memory. Suppose the first leg of its autumn migration is only a few hundred kilometres long and that the bird then lands. By the process of reversal and sideways migrations already described it explores for the best roosting and feeding sites along the latter part of the first leg of the migration, all with no danger of becoming lost. From such a distance, if the worst came to the worst, it could always find its way back to the familiar area around its birth site. However, there is no reason for it to become lost. As it explores for the best stopover sites, it builds a familiar area map in this region. Perhaps the section of the familiar area map around selected stopover sites is not as large as that section around the birth site, but for relatively little effort the map (as opposed to the familiar area) could be 500 km or so across, enough to allow the next migration leg to be up to 1000 km with no danger of being unable to find its way back to the area. This process can then be repeated as many times as is necessary to arrive at the winter range (Chapter 14), whereupon we might expect a section of the familiar area map to be built up as large, if not larger, than that in the breeding area. Spring migration may make use of the familiar area map built up during the autumn migration. Alternatively, it could be an independent process, taking advantage of the winds and habitats available in spring rather than autumn. The further apart it is advantageous for autumn and spring migration routes to be, the larger the familiar area map has to be in the breeding area for the birds to re-encounter their familiar area (Fig. 10.15). I suspect, however, that autumn and spring familiar area maps are usually joined by exploration during the spring migration, knitted together as by a zip-fastener.

Fig. 10.15 The Red-Backed Shrike, *Lanius collurio collurio*: a bird with different autumn and spring migration routes
Solid line, Breeding range; dotted line, non-breeding range; arrows, some known or suspected migrations (short arrows, autumn; long arrows, spring).
[From Baker 1978a), after Dorst]

European
Red-backed shrike
Lanius c. collurio

   In effect, therefore, the familiar area map of a bird that shows long-distance seasonal migration can be envisaged to consist of enlarged sections within the breeding and wintering areas. In between, there is a succession of enlarged, though perhaps smaller, sections around the selected stopover areas (Fig. 10.16). All of these joined by tracts of map only perceived while flying overhead during migration.
   In future years, as an adult, the enlarged areas serve as target areas. The number of such target areas depends on their size relative to their distance apart. Suppose that each target area is only 200 km across, a size that will have required the minimum of sideways migration during the building up of the familiar area in the region. If the bird can orientate in the direction that joins successive target areas to an accuracy of ± 20°, the target areas can be 300 km apart; if ± 10° they can be 600 km apart; if within ± 5°, 1100 km apart; and so on. If we can take a more realistic target area size of about 500 km across, for the same accuracies of orientation they can be 730, 1400, 2900 km apart, respectively. Depending on the size of the sections of familiar area map that are serving as target areas and the accuracy with which birds can work out and then take up the direction joining successive target areas, a journey of 3000 km will require between 11 and 2 such target areas. We are not, therefore, asking a great deal of a bird in suggesting that it can complete its seasonal migrations entirely by route-based navigation and the building up and subsequent use of a familiar area map. Yet even this is the maximum orientation effort that would be required for it assumes that the bird finds its way between successive areas solely by compass orientation without reference to the familiar area map that will

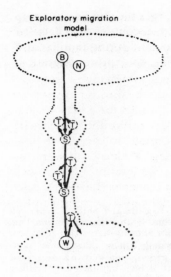

Exploratory migration model

Fig. 10.16 Possible shape of the familiar area map of a seasonally migrant bird
Dotted line, limits of familiar area map. N, natal site; B, potential breeding site located during post-fledging migration; S, stopping site at end of migration leg during first autumn migration; T, suitable transient home ranges for stopover and re-fuelling during subsequent autumn migrations; W, winter home range within the enlarged familiar area map in this region. The spring migration will take place either within the same map or have a different segment of its own.
[Modified from Baker (1978a)]

also exist between successive targets. If it learns, for example, that it can travel between two of its target areas by keeping a stretch of coast in view, 40 km or so to the east and then breaking off to head toward a group of distant hills, 70 km to the south, fly straight over them and keep them to the north for another 70 km, accuracy of orientation does not come into the calculation except under conditions of reduced visibility.

We have not yet considered the effect on this system of wind-drift (by which I mean natural displacement by the wind from some other preferred direction, rather than the situation in which a young bird preferentially travels downwind in order to explore with the greatest economy of effort). If an adult is drifted sideways from its intended direction, it has the entire length of its migration route as a target. Consider a bird with a familiar area map about 1000 km in length, a relatively modest migration distance for a bird. Suppose such a bird is drifted laterally under conditions such that it cannot extend its map as it goes but is forced to employ non-visual route-based navigation. Suppose, as before, that it can do this to within ±10°. Such a bird could afford to be displaced up to 3000 km sideways from its familiar area map before there is a danger of it being unable to find its way back. The chances of an adult bird becoming lost as a result of wind-drift are therefore negligible.

Displacement—release experiments on long-distance migrants have given little evidence by which we can test these suppositions. In part, this is because none have taken route-based navigation into account. In part, also, it is because the familiar area map of the birds concerned is unknown. The famous experiment by Matthews (1968) on Manx Shearwater, *Puffinus puffinus*, across the Atlantic from Wales, in which the bird returned home before the letter saying it had been released, could have been achieved by either route-based or location-based navigation or a combination of the two (Fig. 4.4). In Fig. 10.17 the results are shown of another famous experiment, that by Perdeck (1958) on adults of the Starling, *Sturnus vulgaris*. The map shows the possible limits of the familiar area map of these birds as determined from ringing returns. Again, some combination of route-based and location-based navigation as described in this chapter is adequate to explain the results. Incidentally, the results for adults can be compared with Fig. 9.6 for first-year birds. The latter continued with their exploration following displacement, not returning to their pre-displacement familiar area until the following spring. Young birds seem to treat displacement as free transport during exploration. Sparrows, *Zonotrichia* spp displaced from California to Baton Rouge on the Gulf Coast or to Laurel, Maryland on the East Coast showed the usual pattern. Adults disappeared rapidly from the release point, many of them returning to California the next winter, whereas the juveniles remained in the release area much longer and many returned to it the following winter.

The importance of a large target area can perhaps be inferred from the observation often noted on radar that when land birds have made a sea crossing and strike the coast they fly backwards and forwards along it, many kilometres in either direction. Such behaviour would give a suitably wide section to the familiar area map at a point where lateral drifting is a

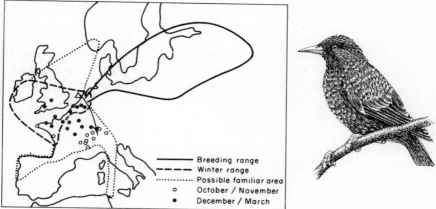

Breeding range
Winter range
Possible familiar area
o   October / November
●   December / March

Fig. 10.17 The behaviour of adult Starlings, *Sturnus vulgaris*, following displacement release in relation to their possible familiar area
All figures relate to birds marked at The Hague (△) in autumn (October/November). Details of the experiment and the results obtained with juveniles have been given in Fig. 9.6. Solid and dashed lines show the breeding and winter ranges of birds that migrate through The Hague. The dotted line shows the limits to the area that Starlings ringed at The Hague have been recaptured, mainly in autumn. This could indicate the area explored by Starlings during their first autumn as juveniles and therefore the limits of their familiar area. Displaced adults stay around the area of release at first (o), only later (●) migrating to the normal winter range.
[From Baker (1978a), using data from Perdeck]

possibility. Another observation also derived from radar observations that lends support to the above picture is that birds are prepared to set off on a migration leg even in relatively bad conditions of mist and low cloud (Evans 1972). Their main concern is to migrate in conditions that indicate conditions are likely to be good at their destination (Nisbet and Drury 1968). The migration can tolerate inaccuracies in initial orientation as long as they are not too great or can be corrected before the bird has travelled too far and as long as the full extent of the familiar area map at the target area can be perceived. Finally, it has been observed that when nocturnal migrants are subjected to apparent wind-drift from their preferred track, they undergo redirected morning flights that take them on the shortest route back to the tract of land they would have travelled had wind-drift not occurred (Gauthreaux 1978).

Description has been strongly biased toward typical temperate land-birds. It needs no modification to be applied to tropical land-migrants (e.g. Fig. 10.18) or shore-birds, nor even to sea-birds, given that the oceans also

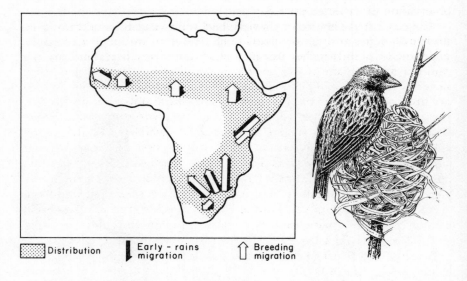

Distribution   Early – rains migration   Breeding migration

Fig. 10.18 Seasonal return migration of Red-Billed Quelea, *Quelea quelea*
During the dry season, the species feeds on seeds that are dry and lying dormant on the ground. As the dry season proceeds, the birds converge on the more suitable habitats. When the rains start, the seeds in the dry season home range germinate and are no longer available as food to Quelea. The flocks then perform an 'early-rains' migration back along the track of the advancing rain front to an area where rain began a sufficient time previously for the plants to have grown and reached the green seed stage. These seeds, plus insect larvae feeding on the plants, are then the main Quelea food source. The return migration back to the dry-season home range proceeds gradually and keeps pace with the movement of the green-seed belt along the track of the rain front. This is a breeding migration. Individual females may raise up to three broods during the course of a single breeding migration back to the dry season home range, each brood at a different point along the migration track.
[From Baker (1978a), after Ward]

provide landmarks as described earlier. On a smaller scale, the main principles of the story can also be applied to terrestrial mammals, bats, reptiles and amphibians, bearing in mind the general definition of a landmark that we are using and the way that different types of landmarks require different forms of location-based navigation (Figs 10.6–8). It may be less obvious how the story can be applied to fish and sea turtles. However, these need only the substitution of Fig. 10.8 for either Fig. 10.6 or 10.7 to account for location-based navigation and the story remains the same: the building up of a familiar area and map by exploration; exploration using route-based and location-based navigation; location-based navigation using the familiar area map as it exists so far; followed finally when adult by movement within the familiar area using learned orientation directions and other self-taught instructions. Although there are few useful data as far as this story is concerned for the mechanisms used by marine fish and sea turtles, some evidence is available for salmon while in freshwater. We shall conclude this chapter by considering this evidence (see reviews by Harden Jones 1968, Stasko 1971, and Hasler and Scholz 1978), and then giving some thought to marine fish and sea turtles.

The case for exploration by young and adult salmon has already been presented. When salmon are raised in a particular stream and then displaced experimentally so that they undergo smolt transformation (Fig. 9.7) in a position downstream of that in which they lived as fry and parr, they nevertheless return *to the area in which they lived as fry and parr*. When they are displaced to a different drainage system before smolt transformation, the salmon return when adult *to the drainage system of release*. When hatchery-raised salmon are not released until several weeks after smolt transformation, adults return *to the drainage system of release* but not preferentially to the stream of release. Finally, when hatchery-raised smolt are released directly into the sea beyond a river's mouth they return *to the river's mouth* as adults but are less likely to enter the river than smolt that were released into the river. Putting all this together, it seems that young salmon collect information as they travel downstream and enter the sea and put this together in a sequential way within their memory.

Sensory impairment experiments on returning adults have shown for many species of salmonids that an intact olfactory sense is of primary importance in 'reading' the map built up when young. Vision may also be involved but is far less important. The most elegant experiments, however, are those carried out by Hasler and his colleagues. Early experiments had shown that Coho Salmon, *O. kisutch*, could discriminate between water from two streams as long as their olfactory sense was intact and as long as the organic fraction of the water was not removed. Moreover, fish trained to recognise water collected from a stream in one season were able to recognise water collected from the same stream in a different season. Recent experiments involved holding Coho Salmon and some trout (*S. trutta* and *S. gairdneri*) during smolt transformation in tanks in which a man–made organic chemical was added to the water. The fish were then marked with fin clips and released in a stream downstream of the tributary that 18 months later, during the spawning migration, was to be scented with the same man–made chemical. A variety of controls were carried out and the evidence seems conclusive: the salmon returned to the stream scented with the same man–made chemical to which they were exposed in a tank during smolt transformation.

For salmon, therefore, a familiar area map, built up during downstream migration when young and based mainly on olfactory landmarks, seems clearly to be implicated, perhaps even proven. Earlier, it was suggested that oceanic fish live within a familiar area (Chaper 4). In this chapter, it was shown that such fish have senses that allow them to recognise a wide range of landmarks, many of which can be detected from great distances. They also have access to various compasses, by which they can carry out route-based navigation and which provide the stable directional axes by which they could orientate their familiar area map within their spatial memory, perhaps along the lines of Fig. 10.8. Probably, no two areas in the oceans have the same chemical signature (Aubert *et al.* 1978). For the moment, we can do no more than suggest that oceanic fish can recognise this fact and use the differences to build up a map in such a way as to base their familiar area and explorations on the same sort of navigational mechanisms that this chapter has discussed for other vertebrates. However, one species of oceanic fish does deserve special consideration, if only because it is perhaps the most enigmatic of all: the European Eel, *Anguilla anguilla.*

All members of the genus *Anguilla* are *catadromous*, spawning in the sea but undergoing the major part of their growth and development in brackish water or freshwater. Fourteen of the fifteen or sixteen species are found in the Indo–Pacific region. All *Anguilla* spp seem to spawn in warm saline water at depths of 400 to 700 m but over very deep water. The eggs are pelagic and develop into leaf-like leptocephali larvae (Fig. 10.19) which then drift with the current. The period of larval life varies considerably from species to species but ends when the leptocephali metamorphose into elvers which swim up estuaries and rivers into freshwater where the eels grow. When the adults mature, often many years later, they migrate back to the sea and are believed to return to their natal area to spawn once and then die.

Although it seems clear that the young larvae are carried from the spawning grounds to their coastal nursery areas on surface ocean currents, nothing is known concerning the route by which adult Eels then close the migration circuit. Once it is full grown, an eel undergoes a series of physiological and physical changes which transform it from the freshwater yellow Eel into the sea-going silver Eel. The latter has enlarged eyes and a retinal pigment characteristic of deep-sea fish. It is assumed that the migration circuit is closed by the adult migrating at some depth in the ocean. However, no adult Eel has ever been caught away from the

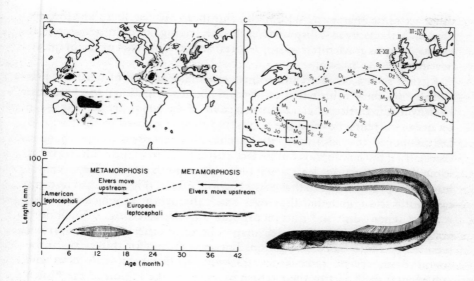

Fig. 10.19 Some postulated migration circuits of eels of the genus *Anguilla*
A. 3 examples of postulated migration circuits. Expected spawning areas are shown by solid black shading and growth and development areas by horizontal hatching. Arrows show surface ocean currents. The North Pacific spawning area is used by a single species, *A. japonica*, whereas the South Pacific spawning area is used by two species: *A. australis* and *A. dieffenbachi*.
B. Growth rate and timing of metamorphosis of eggs and larvae from the southern part of the Sargasso Sea (continuous line) and the northern part (dashed line).
C. Movement of currents from northern and southern parts of Sargasso Sea (M, March; J, June; S, September; D, December) and months of arrival (roman numerals) in different parts of Europe.
[From Baker (1978a), after Meek and Harden Jones]

continental shelf regions of its feeding areas. As a result, there is continuing controversy about the real story of the eel, not least for the deme that spawns in the Atlantic.

The spawning area of the Atlantic deme is assumed to be the Sargasso Sea, though neither spawning adults nor eggs have ever been seen. At the time of writing, an expedition is on its way to the area in an intensive attempt to solve this part of the Eel odyssey. The classical view is that the Atlantic deme consists of two species, the American Eel, *A. rostrata*, which spawns in the southern part of the Sargasso and the European Eel, *A. anguilla*, which spawns in the northern part (Schmidt 1922). Ocean currents from these different parts of the spawning area carry the larvae to their respective feeding areas (Fig. 10.19). When mature, Eels from both sides of the Atlantic return to the Sargasso, American Eels again taking up a more southerly distribution than European Eels. Alternative theories have been advanced, however, ranging from the extreme that only Eels that have matured on the American coasts (Tucker 1959) survive the return journey and that the entire Atlantic deme consists of just one species, to the other extreme that even within the European species there are distinct demes, perhaps even subspecies, each one having its own geographical feeding area (e.g. Mediterranean, Azores) and its own part of the Sargasso Sea in which to spawn.

Against such a confused and controversial background of what the Eels actually do, there is little that can be said about navigational mechanisms. However, the difficulty can be appreciated. If the larvae are carried so far by ocean currents, how do the adults find their way back to the Sargasso Sea (if indeed they do) using the mechanisms described in this chapter, especially if the larvae travel in surface currents whereas the adults travel at some depth? Larvae, at least when older, do seem to show a daily, vertical migration cycle similar to that of other zooplankton (Fig. 4.8), but it seems optimistic to suggest that they ever reach the depths at which they will migrate when adult and thus incorporate the characteristics of such deep water within their familiar area map. The only other possibility would seem to be route-based navigation. Encouragement for this view can be found from recent radio-sonar tracking of European Eels over the continental shelf part of their return journey. In the region of the North Sea, Eels orient to the north-west, the direction from which they came as metamorphosing larvae, known as glass eels. This takes them to the edge of the continental shelf around the northern tip of Scotland rather than through the English Channel (Tesch 1978). Eels transported from the North Sea or Ireland to the Bay of Biscay nevertheless retain their northwesterly orientation, again suggesting route-based navigation rather than location based (Fig. 10.19), though it should be noted that Eels probably also approach the Mediterranean from the north-west. Off Ireland, once they cross the edge of the continental shelf, Eels turn west and south-west, which could suggest the inclusion of the continental shelf on a familiar area map.

In an earlier chapter, the ethological models of fish migration were linked to similar models of sea turtle migration, particularly the Continental Drift model for Green Turtles, *Chelonia mydas*, breeding on Ascension Island. The Continental Drift model has now been dismissed, both on behavioural grounds (Baker 1978a, Chapter 30) and on palaeontological grounds (Gould 1979). Pinnipeds and sea birds prove themselves to be quite adept at finding 'small dots in the ocean' during their explorations when young and to place this small dot on their spatial map in such a way that they can return to it on future occasions. Why should this be beyond the capabilities of sea turtles? In any case, why should the Ascension Island Green Turtles have come into being as a result of continental drift when all over the Pacific there are Green Turtles that are also finding oceanic islands? Here, continental drift cannot have been involved. This is not to say, of course, that continental drift had no effect on the migration circuits of marine animals. Any change in the configuration of land and sea must have some effect on migration circuits. As a general explanation for the existence of oceanic migration circuits, however, Continental Drift cannot seriously be considered. There is as yet no critical evidence. In its absence, however, we should fall back onto our paradigm that, like all other vertebrates, sea turtles will be found to explore, to make assessments and comparisons of the different habitats they encounter, and to navigate by a combination of route-based and location-based techniques, the latter involving a familiar area map, perhaps of the type shown in Fig. 10.8.

This chapter began by noting that in the minds of most people, navigation and long-distance migration were intimately linked. It then proceeded to show that the real link, the real key to an understanding of navigation, lies not with long-distance migration but with exploration. Once this link was appreciated and understood, the relationship between navigation and long-distance movements followed easily and naturally.

The chapter also began by noting that most people felt that navigation was one realm of behaviour in which, without the aid of instruments, humans should feel deficient relative to other animals. The chapter then proceeded along a curious path, placing the final navigational gloss on the view that all vertebrates and at least some invertebrates really do have the cerebral sense of location previously reserved for humans. At the same time, Man also gradually increased in stature. Even when naked of all modern navigational aids, Man emerged as the equal of other animals, deficient in no way. Even the Homing Pigeon may have to accept Man as his navigational equal.

It seems that the harder we look at animal movements, the more we are forced to accept equality among all vertebrates. Some are more spectacular than others, more conspicuous to the human senses, more appealing to the human emotions. At the level of the individual, however, executing its lifetime track against the backdrop of the spatial problems set by a hostile environment, the rules and senses it is using all seem to be variations on the same theme, whether the individual is a human or a fish. Gone, it may seem, are the days of trivial movement, of imprinting, of animal automatons travelling from reflex response to reflex response. Here, instead, are the days of exploration, habitat assessment, familiar areas, and a sense of location. It is time for a new premise on which to base the study of animal migration.

11

A new premise and the search for exceptions

If, like Man, other animals live their lives knowing where they are and where they are going; if they have a cerebral sense of location, live within a familiar area built up by some combination of exploration and social communication, make judgements concerning the best places to live, and find their way around by referring to a familiar area map stored in a spatial memory, then there is no reason to refrain from anthropomorphism. Indeed, we should be creating distinctions where none exist if we did so refrain. If Man solves his spatial problems by thinking, by relating present sensory input to memorised information in such a way as to modify or add to that information or in such a way as to reach a decision, then so too must other animals. We should not be reluctant to accept a cerebral sense of location in other animals simply on the grounds that to do so involves acceptance that they think. On the contrary, if there is evidence that animals solve their spatial problems in the same way as humans, we should use this evidence in forming our conception of the nature of the animal mind and animal behaviour. If we have to attribute human thoughts, feelings, and emotions to other animals in order to understand their behaviour, then so be it; we should be prepared to do so. Equally, if we have to accept that humans may solve many spatial and other problems at a subconscious level, we should be prepared to do that also. Many people regard that making of judgments, assessments and decisions as inextricably connected with full self-consciousness. The Manchester and Barnard Castle expriments on human navigation show clearly that assessments can be made without full reflection on that conscious mind.

Tempting though it is, this is not the place to contemplate those other fields of behavioural ecology, normally associated with Sociobiology—reproductive strategy; sexual, parent − offspring, and other conflicts; territorial behaviour; altruism—to see how these relate to the question of anthropomorphism. As far as migration is concerned, however, the answer seems clear. The paradigm should be that animals do have a cerebral sense of location. They do explore and they do find their way around using the same methods as the naked human. The study of animal migration enters the age of Behavioural Ecology with this as the new premise. An animal is assumed to have a sense of location unless it is proved otherwise.

We began by wondering if there was a natural dividing line between the senses of location of Man and other animals. Having decided that there is not, we then encounter what in many ways are more difficult questions. Are there any animals that, even though they have a sense of location, do not always use it? Are there any animals that simply do not have a sense of location?

We have concluded that all vertebrates and a wide range of invertebrates have a sense of location; but do they always use it? Are there some situations in which a vertebrate, say, embarks on a migration from which it knows there will be little chance of finding its way back to its pre-migration familiar area? We have already been through considerations that apply if the animal uses only non-visual route-based navigation. We should expect most animals, however, to extend their familiar area map as they travel. All we can do is look at a few candidates among, say, mammals and then decide whether such animals ever abandon their sense of location.

The prime example is probably the Scandinavian Lemming, *Lemmus lemmus*, the most notorious but most misunderstood of all mammalian migrants. In order to place the lemming in perspective, however, we should start with others. The adult Muskrat, *Ondatra zibethicus*, given suitable conditions, retains the same monthly home range throughout the year. Muskrat inhabit marshes, streams, rivers, ponds and lakes, and feed mainly on roots, stems, seeds and leaves, but will also take some animal food. Under conditions of drought, the Muskrat moves first to the limits of its home range, about 200 m distant. A temporary roost is established, which may be in the burrows of other species in cornfields and other 'terrestrial' localities. These temporary roosts perhaps serve as a base from which exploration may occur. If a suitable area is not found in which to subsist until the drought breaks and if a return to the local waters remains impossible, the animal may leave the area completely. Under such conditions, marked individuals have been recaptured 35 km from their original home.

Other extreme examples on land are male Eastern Fox Squirrels, *Sciurus niger*, of the United States that have been recaptured 64 km from their original home and a Wolf, *Canis lupus*, that was recovered 670 km from the area in which it had previously been the subject of a radio-tracking study (Camp and Gluckie 1979).

The Sea Otter, *Enhydra lutris*, seems to provide a comparable example. As the species was reduced nearly to extinction during the past 200 years, there now exist within its geographical range (Fig. 11.1) many unoccupied but otherwise favourable habitats. Even when such unoccupied habitats are immediately next to occupied habitats, and even though they receive exploratory migrants from that area, they are not colonised: yet another example of the reluctance of animals to judge a habitat to be a good one unless they have the direct evidence that it supports members of their own species. Such vacant, but otherwise suitable, areas are not colonised until the density in the parent area exceeds about 16 individuals/km^2 (Kenyon

Aleutian
islands

Fig. 11.1 Geographical distribution of the Sea Otter, *Enhydra lutris*, shown by dark areas [From Baker (1978a), after Nishiwaki]

1969). Sea Otters are reluctant to enter water greater than about 60 m in depth. Yet when the deme density in the parent area approaches 16 individuals/km² an adjacent stretch of deep water 15–20 km wide is readily crossed. Nor is 50 km a total barrier at such times and individuals have travelled as far as 100 km across deep water.

Have such long-distance migrants among Sea Otters decided to abandon the familiar area that they built up with their mother and during their own early explorations, to set off instead on a journey into the unknown; or are these movements simply part of the exploration process? Do such Sea Otters retain the option of returning to their parental area, or have they gone beyond the limits of their navigational mechanism? Can they set out knowing that if there really is nothing 'on the other side' of the deep water, they can turn back, having lost nothing but time and energy, or are they forced to keep travelling until they do find somewhere else or until they die in the attempt?

These are the questions we need to be able to answer but cannot, either for the Sea Otter or the Muskrat, Fox Squirrel or Wolf. If the Sea Otter were a seal, we should have no hesitation in assuming that it could return from a mere 100 km (Fig. 6.2), so why should we assume a Sea Otter cannot? Muskrats have shown a 15 per cent homing success following displacement to release points 4 km from their starting home range. Against this, 35 km seems too far for exploration. So, too, does 64 km for a Fox Squirrel. On the other hand, 670 km for a Wolf does not. The familiar area map at the home site will be so large (home range size for the pack in question was 655 km^2; this could suggest a familiar area map of the order of 100 km diameter) and the mobility of the animal is so great, such a distance could easily be part of the normal process of exploration during a wolf's ontogeny. Long-distance travel by itself does not necessarily mean the abandonment of the previous familiar area; the abandonment of the advantages of a sense of location. It does not necessarily indicate non-calculated removal (NCR) migration, a journey into the unknown, but it may do. As yet we cannot tell and can only be subjective. We need more data on the frequency with which animals travel these longer distances and how many ever return, or start to return, to where they began.

A characteristic of all lemmings is that their demes undergo spectacular changes in density. In a flat open area, the migrations associated with these fluctuations take place in all directions. The Scandinavian Lemming, *Lemmus lemmus*, however, lives in a type of habitat that is characterised by strong altitudinal zonation. It is this zonation, combined with extreme fluctuations in deme density, that is responsible for the unique nature of long-distance *L. lemmus* migrations.

The preferred habitat of the species is above the tree-line of the alpine zone of the mountains of Scandinavia. Beneath this level, through the birch and pine wood zones, the suitability of the habitat for Lemmings gradually decreases until an altitude is reached that is unsuitable for colonisation at even low density. As far as Lemmings are concerned, therefore, the alpine zone and, to a decreasing extent, the birch and pine wood zones are islands of suitable habitats separated by large deserts (the lower slopes and the valley bottoms) unsuitable for colonisation.

Deme density increases first at high altitudes. In an expanding deme, successive generations of adolescents are channelled during their explorations to settle further and further down the slopes. A typical sequence of events was described by Curry-Lindahl (1962). Throughout the summer in question, Lemmings occupied their preferred positions on the Scandinavian Fells, though some explorers were always present. Gradually, however, as deme density increased, Lemmings moved down into the willow scrub and birch regions and gradually from there down into the pine forest. A wave of Lemming explorers first invaded the pine forests in May. A second increase appeared in June/July, consisting partly of the offspring of the first wave and partly of downward migrants from higher altitudes. A third major appearance of explorers occurred in the pine forests from August to October.

Such increase in deme density does not necessarily result in the next generation migrating into the valley in large numbers. A qualitative change in migration behaviour seems to come when (and if) the deme density reached in the pine forest exceeds a critical threshold. If it does, Lemmings start to migrate, often in large numbers, down their natal slope, across the valley-bottom, eventually to arrive at some other slope. Although the increase in Lemming density may be more or less synchronous over relatively large areas there is local variation and very often there can be a considerable increase in deme density on the slopes on one side of a valley and much less of an increase, or even a decrease, on the other side (Marsden 1964).

When the threshold for abandoning the natal slope is exceeded the next generation of adolescents together, perhaps, with those adults that fail to hold on to their territory, begin to migrate downslope. Such migrations, at their peak, consist of up to 80 per cent of young individuals. Their course has been described by Bergström (1967).

Time and energy must be at a premium during such long-distance migration and may have more influence than predation risk on the cost of the migration to the animal. Even so, migrating Lemmings retain their predominantly nocturnal activity pattern, though a few continue their migration during the day. Unlike shorter-distance explorers, however, these long-distance migrants take the straightest possible track across unsuitable habitats such as fields and bodies of water. The migration seems almost a race. With such competition, the Lemmings most likely to find somewhere to live will be those that are the first to find a suitable place to settle. As the straightest track across country remains relatively constant from individual to individual, tracks become worn and, being lines of least resistance, are followed by later migrants. Apart from this, Lemmings migrate as individuals, showing no evidence of orientation to other individuals.

We can dispel at once the quaint but unlikely notion that Lemmings are seeking an appropriate place to commit suicide. In order to cross from one side of a valley to another it is usually necessary to cross a water barrier of some kind. In Scandinavia, this can be anything from a stream to a lake or fjord. When large water bodies are perceived at a distance lower down in the valley, Lemmings adjust their orientation in an attempt to avoid having to cross the widest part of the water body. Moreover, when a Lemming arrives at a water barrier it does not start to cross immediately unless it can see the opposite bank. Even then it does not do so unless the water is calm. In fog or wind, therefore, Lemmings may accumulate along the side of a lake during the course of such a night. During the following day they migrate back into the shelter of the forest; crossing is delayed until conditions become favourable. If the weather fails to improve, however, and/or if large numbers accumulate at the water's edge as often happens if they are channeled onto a peninsula, food and shelter may become scarce and many will start to swim even though conditions are still not ideal. Once in the water, a Lemming orientates to topographical features on the opposite side, if these are visible, or to islands.

When crossing ice, Lemmings move at a rate of about 5–6 km/h in spring and 4 km/h in autumn. Once a Lemming sets off across the ice it orientates to the silhouettes of hills across the lake, or to the highest mountain slope. Even so, wide stretches of ice are avoided.

There is no indication here of an animal intent on self-destruction as some film-makers would have us believe. A Lemming gives every impression of an individual, born into a desperate situation, that is doing everything it can to find, in as short a time as possible, somewhere to live. Have they, at the same time abandoned the familiar area that they have so far established? Is their birthplace so bad that these Lemmings will not try to return, no matter what they do or do not find at the end of their journey into the unknown? Intuitively, the feeling must be that they will not. Nevertheless, even during the height of the downslope migration there are individuals travelling in the opposite direction. Are these upslope migrants returning explorers? Have they decided their downslope migration is leading nowhere and that their prospects are better if they return to their original familiar area? Or are they NCR migrants from some other overpopulated slope that are on the final stages of their own search for somewhere to live?

A few months after such a migration, the deme has usually declined and on a mountain slope that so recently supported a steady stream of downslope migrants there will be hardly a Lemming to be seen, though careful observation shows that some are there. Moreover, in one study in which this phase was particularly studied, it was found that whereas the previous autumn a typical downslope migration had occurred, the following spring the few migrants that are to be seen are moving upslope (Bergström 1967). Are these survivors of the previous downslope migration that, having found somewhere to weather the population peak and the crash that inevitably followed, are now returning to their original familiar area? Are they perhaps the offspring of the downslope migrants, performing a re-migration rather than a return? Or are they, as before, hopeful migrants from some distant slope, searching still for somewhere suitable to settle? Until we know the answers to these questions we cannot say whether Lemmings abandon their familiar area mentally as well as physically; whether they are really explorers, judging distant slopes against their own, or whether their prime concern is simply to escape from the conditions of their birthplace.

There is, of course, a human analogy to all this that, with our acceptance of anthropomorphism, we can use to help us sort out our ideas. Throughout human history, at times of war, oppression or famine, humans have vacated their native country in large numbers. Such refugees sometimes walk, forming long straggling columns, creating a visual impression not unlike that of Lemmings. Other refugees may board boats, often totally unsuitable for the journey on which they are about to embark. As in the case of Lemmings, time and opportunity are at such a premium that such people are prepared to take risks that they would not normally consider. Some such refugees may be migrating to a place they have visited previously, or about which they have heard from other people. Many, however, are on a journey into the unknown, their main concern being to escape from the desperate situation in which they find themselves, to find somewhere else—anywhere else, in which to live.

Most refugees are prepared to settle in the first place that will allow them to do so. As soon as they find such a place they begin again to build up a familiar area and map by the usual process. All will remember their place of origin, at least for a time. A few may even attempt to return at a future date. For most, however, such migrations are an abandonment of everything that was previously familiar.

Our image of human refugees, therefore, and one that we can transfer to NCR migrants among Lemmings, Muskrats, and other vertebrates is of an individual that temporarily abandons its sense of location. In any other situation, the ability to return to the previous familiar area and to incorporate any new places found on the existing familiar area map would be at a premium. Now the premium is set by orientation away from the original familiar area, preferably in the direction most likely to lead to a suitable area in which to settle, if there is such a direction. In other words, the sense of location has temporarily been abandoned in favour of a sense of direction.

Butterflies seem to have a sense of location during periods that they are feeding, laying eggs, or holding territories (Chapter 4). Indeed, many of them spend their entire lives in a limited area, alternating between feeding places and others suitable for reproduction (e.g. Wood White, *Leptidea sinapis*; Wiklund 1977). Such butterflies, like Honey Bees, Bumble Bees, some Dragonflies, and probably many other insects, may spend their entire adult life within a familiar area. Other butterflies, however, have a linear range (Fig. 3.4) in which, at intervals, they abandon their current familiar area and set off across country in some preferred compass direction, not establishing another temporary familiar area until they find a place suitable for their present needs.

Whereas Lemmings, humans, other mammals, and perhaps most vertebrates, if they ever abandon their sense of location for a sense of direction, do so only temporarily and perhaps rarely more than once in a lifetime, butterflies with a linear range seem to organise their entire lives around a sense of direction. Our search for exceptions to our paradigm that all animals organise their lives on the basis of a human-type sense of location seems to have led us to the possibility that there is a different way of life: one in which sense of location takes second place to a sense of direction. This way of life, this form of lifetime track, is the subject of the next chapter.

12

A sense of direction: a different way of life

We have to be fairly flexible in discussing whether an animal bases its way of life on a sense of location or of direction. The distinction is convenient, rather than absolute, and should not be pushed too far. After all, animals whose whole life revolves around a sense of location at the same time make considerable use of a sense of direction. We concluded earlier that once an animal had learned that site A was due south of site B, thereafter it would travel between the two, not by navigation, but simply by orienting to the north whenever it left site A for site B. This is making use of a sense of direction, albeit within the framework of a familiar area. Even more pertinent, an animal will, whenever it sets off on an exploratory migration, retain its sense of location by using a sense of direction so that, if necessary it can employ route-based navigation to return to the home site. Route-based navigation can work entirely on a sense of direction. Placing the new site on the familiar area map may not come until after the animal can move between the two by orientation. The clearest example of this is in relation to long-distance bird migrants. First, the young, exploratory, bird migrates in the standard direction using orientation and a sense of direction. Only then are the encountered areas incorporated within the familiar area map by the process of navigation.

We can see that animals with a familiar area make constant use of a sense of direction. For such animals, however, vital though the sense of direction may be, it is a means to an end; the end being a sense of location. Only rarely, as in the NCR migrations of Man, Lemmings and so on, is the sense of direction the end as well as the means. At the end of the last chapter, however, we concluded that some invertebrates, such as those butterflies with a linear range form of lifetime track, organise their entire lives around a sense of direction. These animals, their ways of life, and their sense of direction are the subject of this chapter, but in trying to understand them, we have to beware of two things. On the one hand, we have to keep checking that their way of life really is different from that of the animals considered in earlier chapters. On the other we have to look for attack from those behaviourists that see migrating insects as inanimate particles with next to no control over their own tracks, entirely at the mercy of currents of air and water.

Let us begin by considering those animals that seem to organise their lives around a sense of direction but there is a danger that we are misinterpreting what is really a sense of location. A variety of crustaceans occur on the sandy shores of the world. These range from the large Ghost Crabs (Fig. 3.2) of the tropics to the much smaller sandhoppers, sand fleas, and shore-living isopods that occur in many parts of the world. Most of these crustaceans move up and down the shore on a daily and/or tidal return migration. Their varied orientation mechanisms have been reviewed by Enright (1978). Many roost high up on the shore and emerge, often only at night, to feed lower down on areas of beach uncovered by the receding tide. If dawn breaks before they are immersed by the tide at the burrowing zone, the animals migrate rapidly back up the beach. On such occasions, or when disturbed during the day, learned time-compensated orientation to the Sun is used to locate the upshore direction, as also are local landmarks such as a backshore cliff. At night, learned time-compensated orientation to the Moon can be used, as also can the slope of the substratum and the direction of the wind. In addition, the choice of direction on the landward–seaward axis is seaward on a dry substratum and landward on a moist substratum. On uncompacted moist sand, the animals bury themselves immediately rather than search further. From all of these cues, most of which are learned, the animal evidently knows which zone of the shore it is in and which direction to go in order to get to the zone preferred at each particular time of day or stage in the tidal cycle. The crustacean evidently has a familiar area map of the shore, but perhaps it is not two-dimensional like that of vertebrates and many other invertebrates. Perhaps it is a linear, one-dimensional 'map'; the animal organising its life around the inshore–offshore axis and the sense of direction that this entails rather than a sense of location. However, critical data from horizontal displacement–release experiments along the shore have not been obtained. Perhaps some of these animals do use a two-dimensional sense of location.

On the other side of the tide-line, under the water in the sublittoral zone, are other marine animals that also perform inshore–offshore migrations on a daily, seasonal or ontogenetic basis. Do these have a sense of location or can they, like the crustaceans further up the shore do just as well by relying only on a sense of direction? The North American Hermit Crab, *Pagurus longicarpus*, migrates inshore to shallow water and estuaries in summer but migrates perhaps a few kilometres in autumn to spend the winter buried and inactive in deeper water (Rebach 1978). In this species, there is evidence that some demes learn the direction between summer and winter sites by reference to the Sun. This direction is learned during the inshore migration in spring. In autumn, the learned direction relative to the Sun is reversed and in combination with the slope of the substratum and the chemical composition of the water is sufficient for the crabs to find their way back to wintering grounds 1.6 km away. The directional preference is only found under sunny and partially cloudy skies. There seems, however, to be no compensation for the shift of the Sun's azimuth during the day.

Other demes in a nearby bay seem not to use the Sun. Displacement—
release experiments seem to show that navigation does not occur and that
the migration is achieved primarily by orientation; using a sense of
direction rather than a sense of location.

Other animals that organise return migration cycles around a sense of
direction are members of the zooplankton (Fig. 4.8). Their daily and
seasonal vertical migrations are carried out by referring to light gradients
and gravity: information concerning the direction of up and down. Depth
is recognised from hydrostatic pressure and perhaps temperature. It seems
that zooplankton, like sandy-shore invertebrates, may have a linear 'map'
marked off in zones (of depth in this case) between which they move as
their requirements change during the course of the day and seasons. The
inhabitants of soil, leaf litter and sand, such as pseudoscorpions, worms,
various myriapods, also perform vertical return migrations and may have a
similar linear, zoned 'map' oriented by their sense of up and down.

There is a group of insects that also seems to make use of a linear map
and a sense of direction. Those mayflies (Ephemeroptera), caddisflies
(Trichoptera), stoneflies (Plecoptera) and other insects that feed as larvae in
the running water of streams and rivers invariably become members of the
downstream drift (the name given to the assemblage of organisms, not only
insect larvae, that are caught in plankton nets suspended in streams or rivers
anywhere in the world). Evidently, during larval life, many individuals of
such insects gradually drift downstream. When they emerge as adults,
many of these then fly back upstream in order to reproduce. Such insects
obviously have a sense of upstream/downstream. It has been assumed that a
sense of direction is all they have, but no experiments have been done to see
if, like salmon, they return preferentially to the place in which they
themselves were first deposited by their parents. Is it possible that they have
a two-dimensional sense of location and employ navigation?

The distinction between the way of life of these animals with a one-
dimensional, zoned, 'map' and all those animals we have considered in
earlier chapters with a two- or three-dimensional map is rather subtle. The
difference is much greater when we consider butterflies, the one group that
we can feel confident has a way of life, in those species which show a linear
range, that revolves primarily around a sense of direction and makes little
use of a map. Most information is available for the Small White, *Pieris
rapae*, and this species can be used to illustrate the main features of such a
way of life.

Small Whites feed as larvae on a variety of crucifers but their preferred
foodplants are the various species of cultivated *Brassica*, such as cabbage,
kale and broccoli. As adults they feed on the nectar of a wide range of
flowering plants and roost at night and during cloudy weather on the
underside of almost any broad-leaved plant, such as nettles. Females mate
once soon after adult emergence and thereafter as few times as possible,
rejecting attentive males. Eggs are laid singly. Several hundred are laid

during a lifetime, which in temperate regions such as Britain is about three weeks. The species occurs not only throughout most of Europe and Asia but also through North America, Australia and New Zealand. There are between two and ten generations each year, depending on latitude.

A Small White spends its life travelling across country in a more or less consistent compass direction (Fig. 3.4). As it travels, it alternates between time spent in habitats suitable for feeding, laying eggs, searching for a female, roosting, etc. and time spent crossing unsuitable habitats such as open fields, city centres and roads. While in suitable habitats, the butterfly spends its time flying to and fro, fluttering about, and basking, turning back whenever it crosses over into unsuitable habitats, and generally betraying a sense of location. After it has been in the habitat for a period, its behaviour changes. Instead of turning back when it crosses over into an unsuitable habitat, the butterfly travels straight on, leaving its temporary familiar area behind it. It continues to fly in a more or less straight line until it comes across the next suitable habitat, whereupon the process is repeated. The linear range shown in Fig. 3.4 emerges because each time an individual leaves a suitable habitat to cross an unsuitable habitat it does so in more or less the same compass direction it adopted on previous occasions. Each individual has its own *preferred compass direction*, though as we shall see, different individuals have different preferred directions.

The main difference between butterfly species is in the relative amounts of time individuals spend in suitable and unsuitable habitats. Different species in different areas show the entire gamut from those that spend their entire lives in suitable habitats to those like the Small White that may spend no more than a few minutes to a few hours within each suitable habitat that they encounter—as long, that is, as the Sun is shining. Given sunny conditions, however, stay-time is relatively brief. How brief depends on the size and suitability of the habitat. A really large cabbage field, as long as it also contains flowering plants, may be occupied for more than a day, even if there is continuous sunshine. A small patch of flowers, however, may be visited for only a few minutes. The time spent in each place also varies with age and gender. Small White females migrate across country throughout their adult life, leaving behind them a trail of eggs, the skeleton of their lifetime track. The rate of migration, however, decreases with age (fig. 12.1), a pattern that is fairly general for female insects, though there are exceptions (Baker 1978a). Male insects show a variety of largely unstudied patterns. Male Small and Large Whites (Fig. 12.1) show an increasing rate of migration with age, peaks occurring perhaps at times that few receptive females are to be found.

Small Whites, then, are true nomads. With no fixed home, they gradually make their way across country, feeding, sleeping, and reproducing where they can. Relative to the Small White, human hunter–gatherers and pastoral nomads and other vertebrates, such as the Patas Monkey (Fig. 3.1) are only playing at nomadism. They may not have a fixed home to

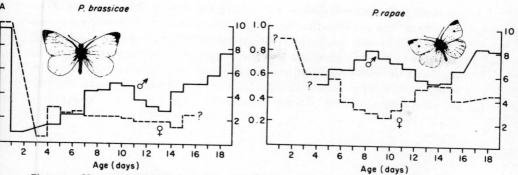

Fig. 12.1 Variation with age in the migration rate of two species of butterflies: the Large White, *Pieris brassicae*, and Small White, *P. rapae*

Average age of the deme studied (in Wiltshire, England) on each day was calculated from observed emergence of adults from pupae collected as larvae during the previous generation and maintained under field conditions. The number of males and females of both species that crossed a field (approx. 1 ha) per hour of sunshine between 11.00 and 13.00 h GMT was counted. So, too, was the mean number in a nearby garden over ten counts spread through the same two hours. Migration incidence in the figure is obtained for any given day by dividing the number crossing the field by the mean number in the garden. Observations made over three generations (July–September 1965; April–October 1966). Question marks indicate an absence of data. Oblique lines joining histograms indicate that no data were obtained for the intervening age(s).
[Simplified from Baker (1978a)]

which to return each night, but they do have a home range within which they return to particular places, albeit at irregular or infrequent intervals.

Why should the Small White, and other insects with a linear range, organise their lives in this way? The answer has to be that no sooner has the butterfly arrived in one habitat than some other habitat elsewhere becomes a better place to be. In many ways it is easy to see why this should be so. The best place to lay eggs or find females is a cabbage field or a garden. The best place to feed, however, is usually elsewhere; not least because the competition for the nectar provided by flowers in cabbage fields is high. The best flowers on which to feed varies with time of day, different flowers producing nectar at different times. Places to sleep and take shelter are abundant. It is easy to see, then, that as the butterfly's requirements change from somewhere to reproduce to somewhere to feed, the best place to be is elsewhere. Having fed, the nearest cabbages are not necessarily the ones last visited. Flowers and cabbages are scattered through the environment in such a way that an individual will do just as well if it keeps travelling into unknown terrain as it would if it navigated its way back to the place it last visited; perhaps even better for there are other, rather more subtle, reasons for seeking somewhere new rather than somewhere familiar.

As a butterfly feeds on a patch of flowers, it may gradually deplete the supply of nectar in that place. Eventually, there must come a point where the efficiency of obtaining nectar in this place is less than it would be if the butterfly went elsewhere, even when the cost (time and energy) of migrating elsewhere is taken into account. This is a situation beloved of the model-makers in behavioural ecology (Charnov 1976, Parker and Stuart 1976, Baker 1978a, Larkin and McFarland 1978). The first two models deal with the special case in which time loss is the only component of migration cost, and the rate of gain of benefit while within a habitat decreases with time. Under these circumstances, the theoretically best time to migrate from a habitat can be shown graphically (Fig. 12.2). The conclusion

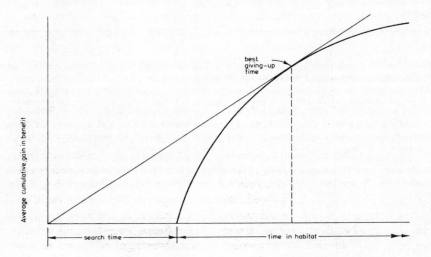

Fig. 12.2 A graphical solution for the optimum stay-time in a habitat when the only cost of travelling between habitats is time lost by searching and when the rate of gain of benefit while in the habitat decreases with time since arrival
[Re-drawn and simplified from Charnov (1976) and Parker and Stuart (1976)]

reached is that the animal should stay in the habitat until the rate of gain of benefit (nectar uptake in the case of the Small White above) falls to a level (called the *marginal value* by Charnov) equal to the average rate for the area as a whole when time spent within and travelling between habitats is taken into account. My own, more general, model (Baker 1978a, Chapters 7–12) is not dependent on such special circumstances and simply states that the animal should leave a habitat when it recognises that the suitability has fallen below the *mean expectation of migration*. For the rest of this chapter we can assume that an animal can recognise when this has happened, but we shall examine the assumption in more detail in Chapter 14.

It is easy to see how, after feeding in a habitat for a while, it may be an advantage to go elsewhere. The same applies, however, for a male searching for a female. If a receptive female is not found, there should come a point, as the male continues to search, at which a female is more likely to be met if the male migrates to another site than if it continues to search in the present site (Parker 1974a). As far as a female laying eggs in a particular site is concerned, there is probably an optimum number to lay in any one place. If she lays all her eggs on a single cabbage plant, they may all survive or all die, depending simply on whether or not a single bird happens to land there. By spreading them out over the countryside, she is much more likely to produce some offspring which survive.

These, and others, are the reasons for continuing to move and for finding new places rather than returning to sites visited previously. The key reason, however, is that suitable sites are so numerous that the distance between them is relatively short compared with the animal's mobility. If the distance is greater, so that the cost of migrating to somewhere new is greater than the cost of returning to a place visited before, then the mean expectation of NCR migration is less likely to be greater than the suitability of the place occupied. The animal may as well stay where it is and take advantage of its sense of location and life within a familiar area. Against this we should note, however, that a short-lived animal does not reap as much benefit from a sense of location as a longer-lived animal.

Exploration and familiarisation are investments in the future. Exploration is as much a journey into the unknown as NCR migration. It is, as we have seen, an inefficient process. The animal in effect tolerates short-term inefficiency as an exchange for the longer-term efficiency and benefits that come from life within a familiar area. If an animal has no long-term future, because it is short-lived, the advantage of familiarisation is less. Nevertheless, it is still an advantage. Exploration is unlikely to be *less* efficient than NCR migration. Everything still rests, therefore, on whether the benefits of familiarity outweigh the disadvantages of return to a site previously exploited. Again, this is a perfect question for the model-makers (see review by Krebs 1978), but we need take it no further. We have found an animal that does have a different way of life, living by its sense of direction, not because it has no sense of location, but because such a sense is not particularly beneficial to it except in the very short term in remembering on which flower it last fed, where it laid its last egg, and so on.

It seems, then, that a way of life centred on a sense of direction can actually be better for some animals than our own way of life; that alternatives have evolved or are adopted, not because the animals concerned are incapable of our way of life but because their way is better for them. Having accepted this, we can now apply ourselves to an appreciation of the mechanisms of such a lifetime track.

Butterflies are not the only group of insects to contain members with a linear range. Many moths, grasshoppers, large flies and probably some dragonflies also seem to organise their lives around a sense of direction, though less is known about these other insects.

Why should an individual have a preferred compass direction? Why travel across country in a more or less straight line? We can see the answer clearly enough while the insect is travelling between habitats. A search path which took an insect back over an area already searched and found to be unsuitable would be inefficient (Baker 1968a, Cody 1971, Charnov *et al.* 1976). The most certain way to avoid doing this is to fly in a more or less straight line. This argument explains why the insect has a straight track while travelling between any two habitats. It also goes a long way toward explaining why the individual should take up the same compass direction when leaving the second habitat as when leaving the first. To do so avoids searching unsuitable terrain already searched. Such behaviour also avoids returning to places in which, for example, eggs have only recently been laid.

We can see, then, why individuals of some species should organise their lives around a preferred compass direction. What environmental cue do they use to maintain their preferred direction? Several species of butterfly

Fig. 12.3 The migration mechanism of butterflies when searching for a habitat the distance of which is unpredictable
A. Small White Butterflies, *Pieris rapae*, were captured during straight line migration across a field, displaced to a different field with different topographical features, and released. Individuals that were flying away from the Sun's azimuth at capture (a) continued to fly at that angle upon release. Individuals that were flying toward the Sun's azimuth at capture (b) continued to fly at that angle upon release.
B. Primary orientation to the Sun's azimuth (p. 171).
The graph shows the mean compass angle of migrating *P. rapae* (●) at different times of day in autumn relative to the Sun's azimuth (solid line). It follows from A and B that an individual migrating towards the Sun's azimuth (● in B(a))flies east of south in the morning and west of south in the afternoon. The result is that the individual, if it flies for several hours, each day describes a semi–circle (B(b)) as it travels across country from one night's roosting site (x) to the next.
C. Secondary orientation to leading lines, such as roads or telephone lines, occurs when the leading line coincides within reasonable tolerance levels, with the preferred Sun orientation direction (p. 171).
D. Reaction to obstacles (p. 171)
E. Reaction to wind (p. 172)
[From Baker (1978a), after Baker and Nielsen]

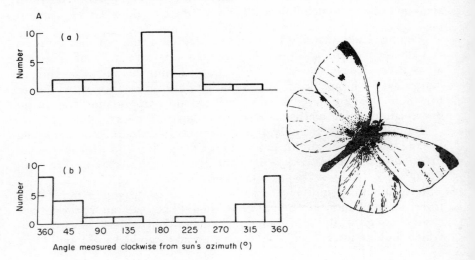

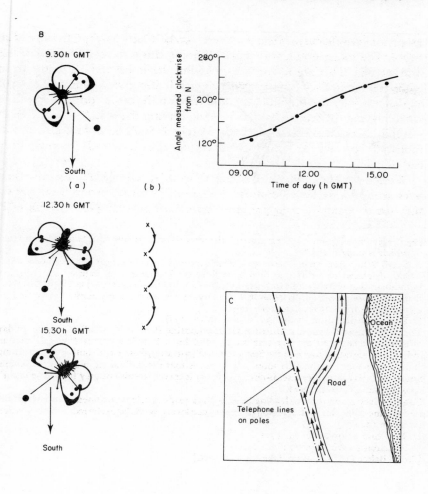

B

9.30 h GMT

South

(a)

(b)

12.30 h GMT

South

15.30 h GMT

South

Angle measured clockwise from N

280°

200°

120°

09.00   12.00   15.00

Time of day (h GMT)

C

Ocean

Road

Telephone lines
on poles

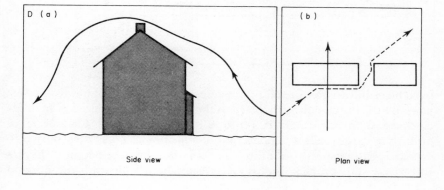

D (a)

Side view

(b)

Plan view

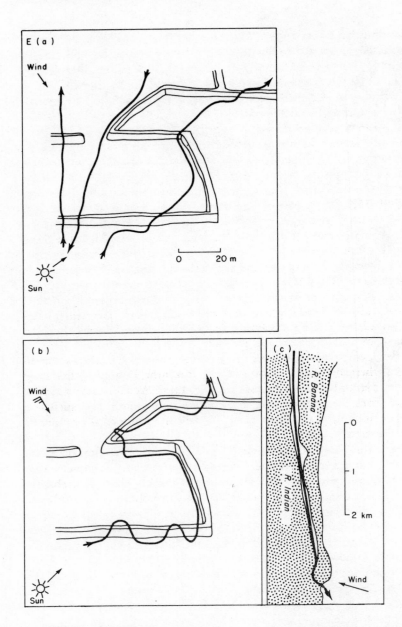

have been shown to orientate with respect to the Sun's azimuth (i.e. the point on the horizon directly beneath the Sun's disk, measured as an angle clockwise from north) (Baker 1968a, b, 1969, Kanz 1977). One species of moth, the Large Yellow Underwing, *Noctua pronuba*, has been shown to orientate with respect to the Moon's azimuth when the Moon is shining and to the stars on clear nights when the Moon is beneath the horizon

(Sotthibandhu and Baker 1979). When neither stars nor Moon are visible, moths can still orientate (Schaefer 1976) by reference to the Earth's magnetic field (Baker and Mather 1982). As in the case of the cues used by navigators, there is a hierarchy of cues for orientation, *least orientation* as well as *least navigation*. Moths prefer the Moon to the stars and prefer the stars to the geomagnetic field, not switching to the next cue down the hierarchy unless forced to do so.

There is one further feature of celestial orientation by most temperate insects such as the Small White and Large Yellow Underwing. Unlike navigators, which are working within fairly fine angular tolerances, these nomads do not compensate for the movement of celestial bodies across the sky, neither Sun (Fig. 12.3), Moon nor stars. A navigator is limited in the distance it can travel during exploration by the accuracy with which it can refer to its compass and for how long the compass is useful. It is essential, therefore, that a navigator, whether human, bird or bee, should compensate for the movement of celestial bodies during the course of the day. A nomad, such as the Small White requires only some means of straightening out its path sufficiently to prevent it from returning to a previously visited area. Orientation at a fixed angle to a celestial body is quite adequate for this (Fig. 12.3) and there is no advantage in spending time learning how to compensate for movement across the sky.

Butterflies avoid wind displacement by flying low and in the shelter of hedges etc. when the wind is strong. When it is light, however, they may fly higher and straight across open spaces (Fig. 12.3). At times orientation is transferred from the primary cue, the Sun, to a secondary cue, such as a leading line, as long as the latter is in roughly the individual's preferred compass direction.

Each individual butterfly, therefore, has a preferred direction based on orientation to some external cue. Different individuals, however, have different preferred directions. Within the population there is a characteristic frequency distribution of preferred directions: a *direction ratio*. Some individual Small Whites prefer to fly toward the Sun's azimuth; some away from it; and some at other angles. Even so, more prefer to fly at some angles than at others: there is a bias within the population toward a particular mean angle. In the Small White, the mean angle relative to the Sun's azimuth during spring and summer in Britain is $159°$ and the direction ratio about this angle is $42:21:21:16$ (i.e. 42 per cent of the population prefer to fly at $159°\pm45°$; 21 per cent at each of the right angles $\pm45°$ to this; and 16 per cent in the opposite direction $\pm45°$). In short, the direction ratio for the Small White is $159°/42:21:21:16$. Different species have different mean angles and different direction ratios (Baker 1978a, Chapter 19).

Not only does the direction ratio very among species, it also varies with time of year. In particular, some time in the autumn there is a major change of direction. In the case of the Small White, the autumn direction ratio relative to the Sun's azimuth is $0°/42:21:21:16$. Note that this mean angle is

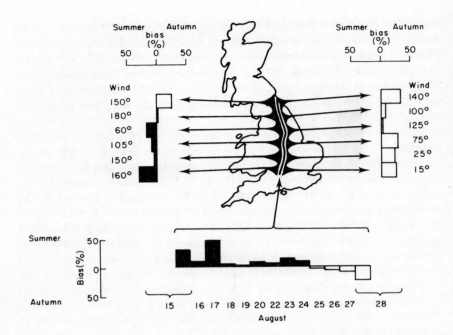

Fig. 12.4 The transition from summer to autumn preferred migration directions by the Small White, *Pieris rapae*, during a single season (1973)

Single-day transects were made along a 480 km N–S line at two stages in the transition period (i.e. 15 August and 28 August). Track directions (relative to the Sun's azimuth) of individuals crossing the road were recorded during each transect. Bias toward summer mean angle (159°) and autumn (0°) was calculated for 80 km lengths of the transect. Bias toward the autumn peak direction is shown by open histograms; summer by solid histograms. Between 15 and 28 August counts were made daily at the southern end of the transect line, mainly at Manningford Bruce in the Vale of Pewsey. Overcast skies prevented migration on 21 August. Number of individuals used in each calculation varies from 30 to 200.
[Simplified from Baker (1978a)]

not the opposite of the mean angle in spring and summer; nor does it need to be for a nomad unlike the seasonal return migrant living within a familiar area and returning to some familiar location. The autumn generation that flies at 0° is not the same as the spring generation that flew at 159° but consists of their offspring. It is a re-migration (Chapter 3). Nevertheless, in the Small White, the changeover from summer to autumn direction occurs in the middle of a generation and takes place over the course of a few days (Fig. 12.4). Each autumn, therefore, a changeover zone moves south across Europe. North of the zone, butterflies are flying south; south of the zone, they are flying north. The zone for the Small White moves south about a fortnight before that for the Clouded Yellow, *Colias croceus* (Fig. 12.5). This zone results from the fact that an individual Small White changes its preferred direction, being more likely to have a southerly preference in autumn after the changeover zone has passed than before. The behaviour is reversible, being influenced by night temperature and cloud cover at dusk and dawn. As these conditions vary from night to night, an individual can change its preferred direction several times in response to the zone passing before eventually settling down to its preferred autumn direction.

The annual re-migration cycle for the Small White in Britain is roughly as follows. An individual of the spring generation gradually migrates across country, taking a preferred direction relative to the Sun's azimuth each time it sets off across unsuitable terrain. This preferred direction probably does not change throughout its adult life. As it travels, it distributes gametes along its lifetime track: females laying eggs, males fertilising females. The length of the lifetime track is probably about 200 km. A member of the autumn generation does the same until it encounters particular conditions of night length and temperature whereupon it adopts a different preferred direction, which is probably more than 90° but less than 180° different from its previous preferred direction. It may oscillate between these two directions for a few days before finally and permanently adopting its preferred autumn direction. As before, the autumn nomad distributes gametes along the length of its dog-leg shaped lifetime track.

The best known of all butterfly migrants, the Monarch, *Danaus plextppus*, of North America, behaves in a manner similar to the Small White but much more spectacularly and introduces us to a new element in the nomadic life of a butterfly. We can again consider our Monarch as an individual. Let us begin with a Monarch of either sex emerging from the pupa in summer in the mid-United States. The chances are that this Monarch will have a preferred direction to the north. Its life will consist, as in the Small White, of a gradual cross-country movement in its preferred direction, orienting with respect to the Sun's azimuth but not compensating for its movement during the day (Kanz 1977, Baker 1978a, p. 433), and distributing gametes along its track as it goes. In Eastern North America many such Monarchs penetrate as far as the Great Lakes before dying.

Consider now an offspring of our first Monarch. After emerging in the Great Lakes region in late July or early August, this individual at first travels slowly, perhaps only 2 km or so a day—half as fast or less than its parent migrated. The preferred direction may still be away from the Sun's azimuth. During this period a great deal of time is spent feeding and basking and moving between suitable sites, building up energy reserves. The young butterfly is non-reproductive and will remain so until the following spring, totally uninterested in sex. Around the middle of August, the chances are that it will change its preferred direction, probably orientating more or less directly toward the Sun's azimuth. It may even

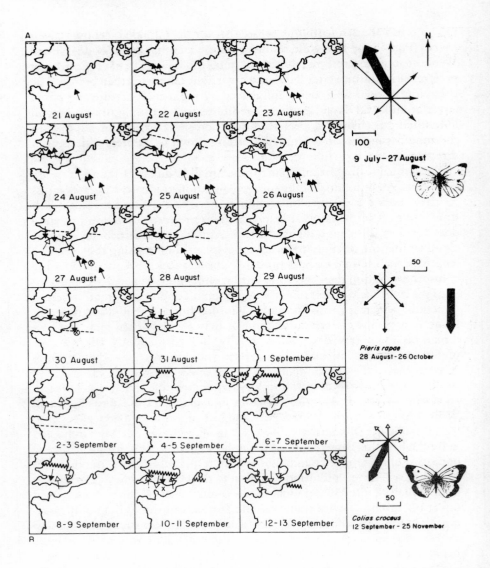

Fig. 12.5 The transition from summer to autumn migration directions by the Small White, *Pieris rapae*, and Clouded Yellow, *Colias croceus*

The right-hand column shows counts of migration direction (arrow length shows the number of individuals flying in that direction) of *P. rapae* in summer and autumn and of *C. croceus* in autumn. Large black arrow shows mean angle. The maps show the way that the expected zones of changeover from summer to autumn migration (dashed line, *P. rapae*; solid line, *C. croceus*) move south across western Europe, that of the Small White, which overwinters as a pupa, advancing approximately two weeks earlier than that of the Clouded Yellow, which overwinters as a larva. The arrows (solid heads = *P. rapae*; open heads = *C. croceus*) show observed mean angles for these two species. ⊗, Indicates no apparent preferred direction for *P. rapae*; x the same for *C. croceus*.

[From Baker (1978a)]

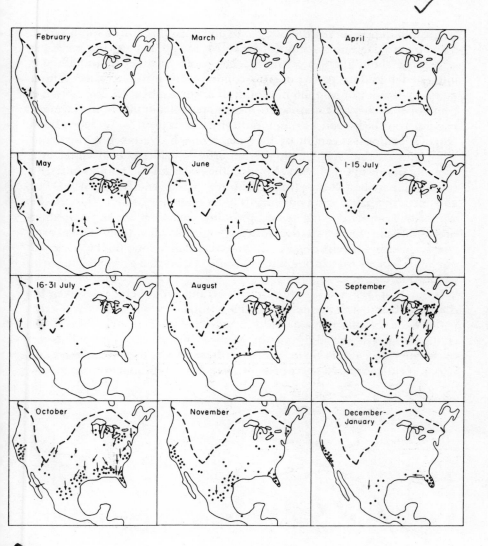

Fig. 12.6 The seasonal return and re-migration of the Monarch
Butterfly, *Danaus plexippus*, in North America
Dashed line, northern limit to distribution in summer; dots,
Monarchs observed flying; arrows, peak migration direction
apparent; X, no peak direction apparent.
[From Baker (1978a) after data from Urquart]

begin to compensate for the Sun's movement so that it flies south rather than toward the Sun's azimuth (Baker 1978a, pp 433−4; but see Kanz 1977). At about the same time its behaviour changes. Although still interested in feeding, its rate of cross-country movement increases. Within 20 to 40 days it can be 2000 km south of its birth-place. By this time it is being overtaken by the southward-moving polar front that heralds winter. Having travelled 2000 km, our butterfly's behaviour now depends on whether or not it is caught by the polar front. If overtaken by the front while still on the Gulf Coast, the butterfly is liable to overwinter as a member of one of the clusters that form the beautiful and famous butterfly trees. These are best known from California (Tuskes and Brower 1978) but also occur in Florida and Louisiana. If it does join one of these clusters, the winter will be spent taking advantage of fine days by flying out to feeding sites and then returning to the cluster at night or when the weather turns cold once more. Navigation experiments have apparently not been carried out but it is likely that during winter a sense of direction gives way to sense of location.

If our Monarch manages to reach Texas and avoid being overtaken by the polar front, it continues its southward journey but at a slower rate, perhaps only 5 km or so a day (Fig. 12.7). Nevertheless, many travel fast enough to reach the highlands of Mexico (Urquhart 1976) and enter overwintering clusters there similar to those of California and the Gulf Coast. Our Monarch becomes interested in reproduction and first

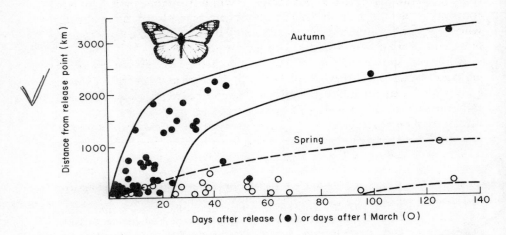

Fig. 12.7 Migration rate of tagged Monarch Butterflies, *Danaus plexippus,* in spring and autumn

The graph shows the distance from their release point that marked Monarchs have been recaptured at varying time intervals after release in California in spring (o) and in the Great Lakes Region in autumn (●). Migration rate during a 20−30 day period in autumn is different from, and greater than, migration rate at any other time of year. Some autumn individuals seem not to show the faster rate of migration.
[From Baker (1978a), after data by Urquart]

copulates perhaps as early as January. The first permanent departure from roosts does not occur, however, until the beginning of March, when the preferred direction is slightly more likely to be north or away from the Sun's azimuth than any other. Spring is a time of slow, cross-country travel, very similar to that of the Small White, travelling less than 15 km/day. Oviposition begins in March and as our Monarch travels from its winter site it distributes gametes along its track in the usual way, moving from feeding site to reproduction site to roosting site, and so on. As the Monarch is longer-lived in spring than Small Whites, many living perhaps until July, our individual will travel more than the 200 km or so typical of the smaller species before dying—perhaps nearly 1000 km (Fig. 12.7). Even so, it is unlikely that it will ever see the Great Lakes again, but will die somewhere en route. It is certainly unlikely that it will be one of the Monarchs to appear in the Great Lakes region by the end of May. It is still possible that these are individuals which overwintered in the region without making the great trek to the south (Baker 1978a, Chapter 19).

The southward autumn migration of Monarchs seems to be a race to keep ahead of the polar front that brings winter to North America. Like Lemmings, straightened out movement is at a premium. Leisurely cross-country movement, concentrating on detecting any suitable habitat that comes into sensory range, gives way to intense southward migration, which takes priority over everything but feeding and roosting. Everything is geared toward speeding movement to the south, the butterflies even rising high into the air to take advantage of winds from the north. This is in contrast to the situation in Small Whites, or even Monarchs in spring, where everything is geared toward detecting the next suitable place to feed or reproduce that comes within sensory range.

Behaviour similar to that of the Monarch may be shown by other temperate-zone butterflies (Red Admiral, *Vanessa atalanta*, and Painted Lady, *Cynthia cardui*, of Europe and North America are the main examples) and moths (such as the Silver-Y, *Plusia gamma*). It may also be shown by sub-tropical moths (Fig. 12.8). It is certainly shown by tropical butterflies in areas where rainfall is seasonal.

The main feature of the arid grasslands and semi-desert regions of the world is the seasonality of the rainfall. Zones of rain move back and forth over these regions bringing with them shifting areas of water, vegetation, and animals (see Figs 3.9 and 10.18). The animals of such regions have only two choices: to remain in one place, like the plants, and wait for the rains to come to them; or to be nomads, forever chasing the retreating storms. Many butterflies seem to adopt the latter alternative, probably having large migration circuits and several generations elapsing before the annual circuit is complete (see Fig. 3.9).

As one generation of such a butterfly emerges, it has to travel along the circuit to catch up with the retreating rain and vegetation. For any individual butterfly, finding itself at a particular position on the circuit, the

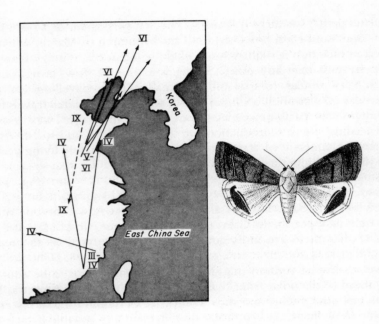

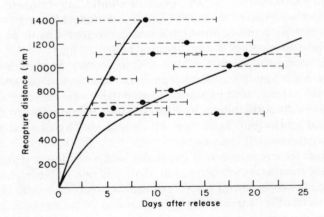

Fig. 12.8 The migration of the Oriental Armyworm Moth, *Pseudaletia separata,* in China
The map shows the results of a mark – release – recapture experiment on *P. separata.*
Arrows join the positions of release and recapture. Solid lines, spring and summer migrants;
dashed line, an autumn migrant. Roman numerals indicate the months of release and
recapture.

distance and direction of rain are more or less predictable. The challenge is to get there as quickly and economically as possible: there is no point in searching for a place to reproduce until then. This challenge is the same as for the autumn Monarchs. The solution also seems to be the same, except that nearly all individuals adopt the same direction. There is no point in flying in other directions. Once adult, the trek cannot be avoided. The direction ratio is virtually /100:0:0:0, all individuals flying in more or less the same direction. Stopping only to feed and sleep, a typical arid-region butterfly migrant flies low against head-winds but rises high to take advantage of tail-winds and uses its sense of direction to wring every assistance from its environment. We do not know the basis of a tropical butterfly's sense of direction, but if orientation to the Sun is used, then compensation is also made for its movement across the sky. Tropical butterflies maintain a consistent compass direction at all times of day.

Butterflies are not the best-known of tropical migrants. This distinction belongs to the locusts: short-horned grasshoppers, adapted to life in arid environments, characterised by at least occasional aggregation into vast swarms that steamroller over the countryside, demolishing all vegetation in their path. Locusts are the Jekyll and Hyde of animal migrants. If they develop when young at low deme densities, they mature into perfectly ordinary short-horned grasshoppers, lead a solitary existence, are cryptically coloured, and migrate at night. If they develop when young at high deme densities, they form hopper bands, synchronise their development, and mature into locusts. They are then highly gregarious, brightly coloured in yellows and reds, relatively long-winged, and migrate by day in vast swarms. For a time these two 'faces' were thought to be different species. The classical work of Uvarov (1921), however, proved them to be different phases (*solitaria* and *gregaria*) of the same species. This phase-transformation has evolved independently several times, all the arid regions of the world having their particular species of locust.

Where data are available all locusts have migration circuits of the type described for arid-region butterflies. The best studied is the Desert Locust, *Schistocerca gregaria*, of Africa and Asia (Fig. 3.9). The interpretation given here is based on a previous analysis (Baker 1978a, Chapter 19).

When a Desert Locust becomes adult and capable of flight, the nearest rainfall will be some distance away, along the next leg of the migration circuit, in a relatively predictable direction. The Locust is non-reproduct-

---

The graph shows variation in the distance of recapture with days after release. Because each batch of moths was released over a period of several days, the precise time interval between release and recapture cannot be shown. The dot indicates the time interval between the median release date of the batch and recapture of the individual concerned. The dashed bar shows the range of possible intervals between release and recapture for each individual. The two solid lines delimit most of the observed migration rates.
[From Baker (1978a) after data and map by Li, Wong and Woo]

ive when it first becomes adult and will remain in this reproductive diapause until it encounters rain. If the locust is phase *solitaria* it migrates at night, as an individual, paying little, if any, attention to the behaviour of other conspecifics. Like a moth, it travels across country in its preferred compass direction. Perhaps it travels low, or not at all, against a head-wind, but rises up into the air to take full advantage of any assistance offered by a tail-wind. In some species, such as the Australian Plague Locust, *Chortoicetes terminifera*, whole sections of the migration circuit may be completed by generations in phase *solitaria*.

If our Desert Locust is phase *gregaria*, it emerges as an adult and starts to fly amidst hordes of conspecifics. Much of its behaviour is geared to the behaviour of those around it and the swarm will stay together for most of their lives. All members of the swarm have roughly the same preferred compass direction, though occasionally a swarm divides into two, each part going off in a slightly different direction. Once our young but adult locust is able to fly strongly, it sets off with its thousands or millions of conspecifics to complete the next leg of the migration circuit, to catch up with the retreating rain and the new vegetation that appears in its wake. With a tail-wind our locust rises high into the air, taking full advantage of the wind to

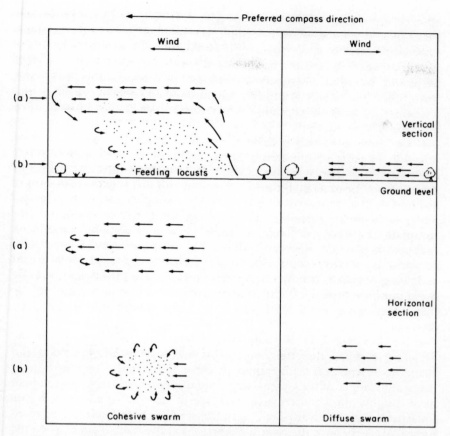

Fig. 12.9 The structure of cohesive and diffuse swarms of the Desert Locust, *Schistocerca gregaria*, and a postulated relationship between preferred compass direction and wind direction
[From Baker (1978a). Swarm structure from descriptions by Waloff. Photo courtesy of C. Ashall and COPR]

travel as fast and economically as possible in its preferred direction. From time to time it settles on the ground beneath the swarm in order to feed on such vegetation as is encountered, rising into the air once more when satiated. While flying high, hundreds, perhaps thousands, of metres above the ground, our locust travels faster than those beneath it, particularly, of course, those feeding on the ground. Eventually the locust emerges from the front of the swarm and can see no other locusts beneath it. Staying as a member of the swarm seems to be a priority, perhaps because of the protection this offers against the hordes of predatory birds and other animals that feed on locusts, perhaps because it ensures it will be amongst conspecifics and can immediately start to reproduce when rain is encountered. As a result of this priority, our locust, as it emerges from the front of the swarm, drops or flies toward the ground and orientates back into the swarm. This may be when it settles to feed. While it does so, it observes the passage of the swarm overhead. If the tail edge of the swarm passes over while the locust is feeding, it immediately flies up, orientates toward the swarm, and rejoins. If it ever emerges from the side of the swarm it again orientates back into the swarm. As a result of such behaviour by all its members, the swarm is cohesive (Fig. 12.9) and steamrollers over the countryside, vacuuming up food by virtue of the locusts settled beneath it, gaining motive force from the oriented flight of the locusts in its upper levels.

If the downwind and preferred compass directions coincide, as they often do in parts of the Desert Locust range, our locust can take full advantage of assistance from the wind. If downwind direction begins to deviate from the preferred compass direction, our locust when in the upper levels of a cohesive swarm adjusts its orientation so that its track is on the preferred compass direction side of downwind. Compensation is not complete, being only $10°$ different from downwind even when preferred compass and wind directions differ by $70°$, but may be the best compromise between continuing to take advantage of wind displacement and being deviated from the straightest course toward the nearest rain by the minimum amount. In other words, degree of compensation may be that which gives the largest vector per unit energy in the preferred direction.

If the downwind direction deviates by much more than $90°$ from the preferred compass direction, our locust flies low against the wind and abandons orientation with respect to other individuals, there no longer being any danger of the swarm becoming dispersed because all individuals have roughly the same cross-country speed; the swarm adopts a diffuse formation with no clear edge (Fig. 12.9). Diffuse formations are adopted most frequently once the swarm has arrived at the rainfall zone. Then the locusts may abandon their preferred compass direction to fly instead toward high ground (Fig. 12.10), where rain may be more likely, or toward distant rain or rain clouds that are in sight.

Large, strong-flying insects with a linear range, such as many butterflies, large moths, grasshoppers and locusts, have so far been presented as animals that use a sense of location only temporarily: instead, they have a strong sense of direction. The image is the classical one (Williams 1930, 1958). It was first challenged in an important paper by Rainey (1951) with respect to the Desert Locust. Rainey argued that locusts did not have a sense of direction and that their primary orientation was toward the swarm, which as a result had many of the properties of a balloon and was rolled across the countryside by the wind. As in the tropics the downwind direction is frequently toward the zone of rainfall, it seemed that locusts had no need of a sense of direction: they would arrive at a suitable destination simply by delegating the determination of direction to the wind. Within the general ethological climate of the time and the search for ways of conceiving animals as automatons, the view spread and attempts were made to explain

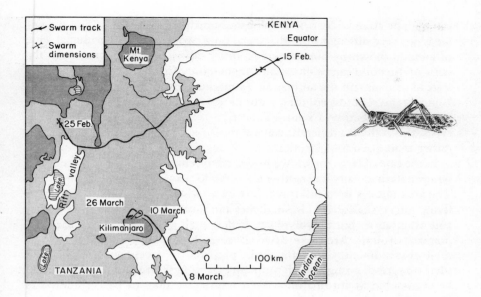

Fig. 12.10 Tracks of two swarms of Desert Locust, *Schistocerca gregaria,* in relation to topography
[From Baker (1978a), after Rainey]

all insect movement in terms of downwind displacement. With the publication of the major essay on insect migration by Johnson (1969), this view became that of the entomological establishment. The 1970s saw a gradual erosion of this view and a re-establishment of the image of animals with a sense of direction. I have reviewed the arguments of both sides (Baker 1978a, Chapter 19). Since that review was written (in 1974), further information has appeared from radar studies of butterflies, large moths, and grasshoppers (Schaefer 1976, Riley and Reynolds 1979) and the controversy would now seem to be more or less over. Visual and radar studies, where they provide critical evidence, all show that butterflies, the larger moths, and grasshoppers either maintain their preferred direction irrespective of wind direction or at least compromise between their preferred and the downwind direction. This, combined with the evidence for orientation to celestial cues, makes the importance of their sense of direction beyond doubt. It is, however, too soon to claim that this is also true for gregarious locusts and I suspect that debate will continue for some time yet Draper (1980), Baker *et al.* (1981).

It seems that there are two ways in which an animal can organise its life around a sense of direction rather than a sense of location. It can use its sense of direction to constrict its environment into a linear, one-dimensional 'map' in which everything is reduced to zones or positions on a single axis. The immediate surroundings are then used to identify position on this 'map' and to work out the direction of the preferred position. This position can then be attained by orienting in the appropriate direction. This way of life is not very different from that based on a two- or three-dimensional sense of location shown by humans and so many other animals. Alternatively, the sense of direction can be dissociated from sense of location entirely and used instead to maintain the animal on a straightened-out track, the prime aim being to travel across country without retracing its steps. Except that such animals often betray a sense of location while within preferred habitat types, this really is a different way of life—one based on a sense of direction rather than a sense of location.

In the case of large insects we managed to rebuff the idea that there could be animals that really had neither a sense of location nor a sense of direction. The idea returns with full force, however, when we turn to small, weak-flying insects such as aphids, small flies and moths, leafhoppers and the like, and other aerial but non-flying travellers such as spiders, caterpillars and pseudoscorpions. Are there, after all, animals that abandon themselves to their environment, like plant seeds, prepared to accept whatever comes their way, responding to each moment as it comes? If so, are such animals to be considered as automatons? Even worse, are they to be considered as inanimate particles, no different from wind- or water-borne plant seeds or particles of dust? These would seem to be the final questions in our search for different ways of life, different solutions to the spatial and temporal problems set by the environment—final, that is, except for the big one, the one that is really behind all of the other questions. We can save that, however, until the end of the next chapter.

13

The automaton: life without a past

Animals that live within a familiar area are forever relating present situations and circumstances to past events and experiences. Even those animals, such as sandhoppers, with a linear map which they travel simply by using a sense of direction, are nevertheless continually referring back to the past. Butterflies, moths, grasshoppers and the other large insects with a linear range, however, are continually on the move across country. Their learning seems only to begin when they enter a suitable habitat. Then they may begin to betray a sense of location, referring to their immediate past, the time since they arrived. As soon as they leave there seems no reason for them to remember that place; they are unlikely to return to make use of such memories. All they need from the past is a continuation of their preferred compass direction. If there were an animal with no sense of location while in its normal habitat and that showed no sense of direction during cross-country movement, such an animal would be one without reference to its past in determining its future movements. Such an animal would be an automaton, generating requirements within itself according to some internal programme, matching the environment to these requirements, responding accordingly. Is there such an animal?

Whether an aphid has wings when it becomes adult depends on many factors: time of year, photoperiod, temperature, gender, species, and how crowded it was when younger or perhaps even how crowded was its mother. Some aphids live on the same species of plant all year. Others alternate between different hosts at different times of year (Fig. 13.1). Without wings, an aphid cannot migrate by flight; even with wings it may not. The following picture of what happens when an aphid does decide to migrate by flight is based on a previous analysis (Baker 1978a, Chapter 19). It differs from the classical view (Johnson 1969) in a number of ways but these are relatively unimportant for the purposes of this chapter.

When an aphid flies into the air from its natal plant, its first response is to fly upwards until it is above nearby vegetation, thereby avoiding the danger of being blown into spiders' webs and other death-traps. Within 15 to 30 seconds, it begins to search for somewhere to settle. Its decision whether to stay and search in the vicinity of the plant on which it was born may have been made before it took flight. Alternatively, it may depend on how many other aphids are in the air or perhaps on how many aphids it

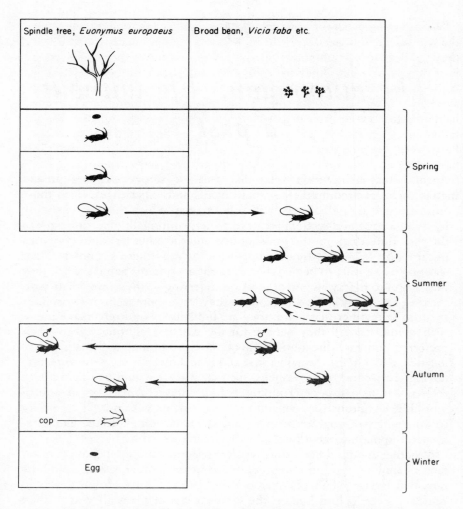

Fig. 13.1 Life-history of a host-alternating aphid: the Black Bean Aphid, *Aphis fabae*
Left-hand column, part of life-history spent on Spindle Tree; right-hand column, part spent
on secondary hosts. All aphids shown are females except where indicated. Presence and
absence of wings are as indicated but in summer the proportion of winged to wingless forms
varies considerably and is not necessarily 1:1 as shown. Females shown in black are
parthenogenetic and viviparous. The generation shown in white is oviparous and copulates
with males as shown. Horizontal lines separate generations except in summer when number
of generations is variable. Continuous arrows indicate obligatory change of host; dashed
arrows indicate migrations to new secondary plants in response to local conditions.
[From Baker (1978a)]

encounters if it does settle on nearby shoots and leaves. If it decides to travel
away from its natal patch of plants and search for a new site elsewhere, it
flies up into the air until it is above its visual horizon. An aphid's power of
flight is so weak that once above its visual horizon it has no alternative but
to delegate the direction of its displacement to the wind.

We can assume that for most aphids the advantage lies with travelling on the wind at a height just above the highest vegetation, usually trees, so that they can drop into a potentially suitable habitat as soon as they come across it. If they fly too high, they have less chance of dropping into the habitat before being carried too far beyond it to be able to battle their way back against the wind. If they fly too low, they have less time to perceive the habitat before being carried past and perhaps of having a greater risk of falling prey to spiders and other predators present in the vegetation. However, once an insect as weak-flying as an aphid is above tree height, it has little control over its track. In particular, convective currents are liable to carry the animal high into the air, perhaps hundreds or even thousands of metres above ground level (Fig. 13.2). We should expect that when this

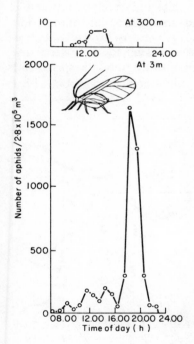

Fig. 13.2 Hourly densities of aphids at two heights above ground level at Cardington, England, on 8 August 1955
[From Baker (1978a), after Johnson]

happens an aphid should attempt to return to a height just above the vegetation as quickly as possible, either by simply ceasing to fly or by flying actively downwards (Thomas *et al.* 1977).

As soon as the aphid encounters a potentially suitable habitat, it drops down from its free transport and begins a detailed search of the area for somewhere to settle. Even once a leaf or shoot has been chosen, it may be vacated several hours later or even the next day. Subsequent journeys may be shorter and in the same general area (e.g. Taimr and Kríz 1978). Eventually, the aphid settles permanently and often spends the rest of its life on the plant selected. In some species, the wing muscles atrophy.

The differences between the aphid's way of life and that described for the Small White do not seem all that great. The individual controls its own take-off and landing; it may even show a sense of location during the final stages of search for somewhere to settle, though this seems not to have been investigated. During migration it does what it can to control the height at which it is transported. The only thing it does not do, it seems, is control the direction of its migration. Indeed, there seems no reason for most temperate zone aphids or similar weak-flying insects to try to control their migration direction. Most such animals migrate by flight only once, settling more or less permanently at their destination. During the course of a single brief migratory flight the wind, even in temperate regions, is likely to give the insect a straight enough track not to conflict with the requirement of the avoidance of retracing its steps and searching unsuitable areas more than once. There is, then, little advantage in a sense of direction for such brief, weak-flying migrants. Some small insects, however, such as the Beet Leafhopper (Fig. 13.3), perform a north–south seasonal re-migration. Perhaps they do select a south wind for migration in spring and a north wind in autumn; perhaps they do combine aeolian transport with a sense of direction.

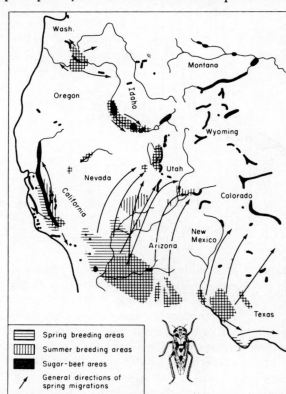

Fig. 13.3 The migrations of the Beet Leafhopper, *Circulifer tenellus* [From Baker (1978a), after Johnson]

Aphids and other small insects, therefore, may or may not make use of a sense of direction. Perhaps, during the search for a final place to settle, they even show exploration, site ranking, and other behaviour associated with a sense of location. On the other hand, perhaps they respond only to the situation of the moment, taking no account of memories of past locations. Take off, flight control, and settling could all be achieved by an animal that makes no reference to the past: an automaton. However, the evidence is not yet adequate to dismiss our new paradigm that all animals have a sense of location and direction. Aphids and other small insects could be our automatons, but the case has not finally been proven.

The same can be said for those other aerial travellers among arthropods, the gossamer or aeronaut spiders. These certainly control take-off, climbing to the top of a plant, paying out a length of silk until it is taken up

Fig. 13.4 Aeronaut spider before take-off
[From Baker (1978a)]

by the wind, and then releasing its hold. Such a spider may even control the height of its displacement and where it settles by appropriate shortening and lengthening of the silk by which it is being ballooned along. The situation seems little different from that of an aphid or other small flying insect.

The situation may be different, however, for perhaps the most unlikely of all members of the aerial plankton: caterpillars. Some small larvae may be caught at great heights in the air. These could be accidental migrants, wrested from their host plant by a sudden upcurrent of air to be deposited wherever the wind chances. On the other hand, most small caterpillars spin a silken thread as they crawl over their host plant; this is their lifeline by which they hang and crawl back up should they ever fall off. Perhaps these behave in just the same way as gossamer spiders. The fact that many of these young caterpillars are very hairy, which would aid wind transport, and belong to species the adult females of which are wingless, suggests that the migration is often initiated actively rather than by accident. Whether, like aphids and perhaps spiders they can control the height of displacement and the moment of dropping out of the air we can only, for the moment, leave to the imagination.

Pseudoscorpions and some mites migrate through the air from place to place by phoresy, often attaching themselves to the leg, or some other part, of an insect. Take-off is very much under the control of the migrant. Attaching itself to an insect is comparable to a human boarding a plane, ship or horse. There the analogy ends. Unlike a human, the pseudoscorpion is travelling to an unknown destination. If it has boarded the right insect, however, it is likely that the next place the vehicle stops will be a suitable one for the passenger. Clearly, no sense of direction is involved in the pseudoscorpion's journey. This does not mean, of course, that the animal has no sense of direction or location while within its preferred habitat.

Turning our attention away from the air, we find other animals, living along the edge of the tide on the sea shore, that at first sight also seem to have no sense of direction or location. A wide variety of animals occur here that maintain a more or less constant position relative to the water's edge, moving up and down the shore with the tide. Bivalve (e.g. *Donax denticulatus*) and prosobranch (e.g. *Hydrobia ulva*) molluscs, mole crabs (Fig. 13.5) and amphipods (e.g. *Synchelidium*; Enright 1978) all show this type of behaviour. Position is maintained by a combination of digging in (to maintain position on the beach) and swimming (to move up or down the beach), along with discriminating between incoming waves and backwash and an appreciation of up and down using light direction or gravity. As described in a recent review (Enright 1978), such animals could easily be the automatons for which we are searching, each individual responding to the cues of the moment to meet requirements engendered by an endogenous programme under the control of a biological clock entrained by the tides. However, we have to be careful. Apart from the

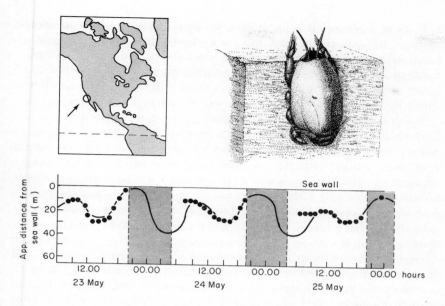

Fig. 13.5 The tidal migration of the Mole Crab, *Emerita analoga,* on the coast of California Solid line, approximate position of edge of tide; dots, observed positions of an aggregation of mole crabs over the same period.; stippled areas, hours of darkness. The aggregation moves up and down the beach at the edge of the water. When the tide recedes toward its lowest point for the day, the crabs stop at the level of the maximum of the next high tide. [From Baker (1978a), after Efford]

sense of direction their behaviour involves (up versus down), most such animals tend to settle out from the tide as it recedes, there to await the next incoming tide. Moreover, the position on the shore at which they settle to wait is related to the height of the next tide (Fig. 13.5). Such behaviour could betray a learned sense of location, at least on a linear map like that possessed by other shore animals. It could, but it might not. These animals could be our automatons, our animals without a past, but again the case has not been proven.

We are forced in the end to return to the barnacle larva on which we pinned our hopes as early as Chapter 4. Does a barnacle larva searching for somewhere to settle avoid places visited previously if they were unsuitable? Does it recognise a place as one having been visited before? Or does it, like an automaton, employ a pre-programmed search pattern and settle in the first site it encounters that is above its present threshold for settling? Even if a barnacle larva is not quite the automaton for which we are searching, what about protozoans? Do they meet all the requirements of an organism without a past? Or do we really have to look to plants as the only organisms without a sense of location (Baker 1980b). Once a plant seed has left its parent, either jettisoned, or carried away by wind, water or gravity, its final resting place as an individual is determined entirely by the environment. If it lands in a place that is unsuitable, it can do nothing about it, except, perhaps, to lie dormant until the place improves. It cannot compare its location with some other location: it cannot visit any other location. It could be argued that once germinated a plant has a sense of direction, of up and down, towards the light and away from the light, but in no useful way can it be described as having a sense of location.

Is this eventually where we have to draw the line: between plants and animals. Or are there some animals that have no trace of a sense of location? Maybe there are, but the case has not been proven. Our new paradigm remains unviolated. Intuitively, one feels that there must be animals without a cerebral sense of location, that at no time in their lives explore or compare habitats or take any account of previous positions along their lifetime track. Again, however, we can wonder whether this is intuition or prejudice. What is certain is that, if there is a dividing line between animals with a cerebral sense of location and animals without, it does not lie between Man and other animals; nor even between vertebrates and invertebrates. It may even be that it lies between animals and plants.

Surely, though, this cannot be the final conclusion? Suppose that all animals do have a cerebral sense of location like Man. Surely this cannot mean, as we have implied, that at some stage in their life all animals know where they are and where they are going and make judgements and comparisons between habitats; in other words that they think? Surely there still has to be some qualitative difference between the human, or at least the vertebrate, sense of location and that of invertebrates? Surely, the final argument would be that this difference has to be something to do with the relative involvement of instinct and learning, a relationship that has carefully been avoided in all previous chapters? This is the big question; the one that finally focuses our attention on the real nature of the sense of location; the one that inexorably leads us into the debate at the centre of behavioural ecology and sociobiology. What relevance has the study of the behaviour of other animals to the understanding of human behaviour? It is time for the study of migration finally to enter the real fray.

14

Learning and instinct

This book began with the image of an animal as an individual travelling through life, attempting to solve the spatial and temporal problems created for it by a hostile environment. The individual's weapons in solving these problems were seen as its innate predispositions, the product of generations of natural selection, which are then finely tuned by personal experience to suit the particular environment into which the individual is born.

Since creating this image, we have steadfastly ignored these innate predispositions, concentrating instead on the way they are modified by personal experience. This was intentional: an attempt to counter the legacy from the past by going as far as possible in the opposite direction. The legacy received by the study of migration from the age of ethology was an attempt to explain all migration behaviour in terms, not even of innate *predispositions*, but inflexible innate *programmes*. The obvious response was first to see how much of migration behaviour can be explained in terms of learning. A great deal, seems to be the answer. Behavioural ecologists, however, realise that even the rules for learning must be under the influence of innate predispositions. This is where the conflict begins with the proponents of human uniqueness. Many sociologists and psychologists— indeed, many biologists—see a human individual as a blank page upon which experiences before, during, and after birth write down the predispositions, predilections and prejudices that the individual will show in later life.

At first sight, the picture of migration behaviour given in this book, at least for those animals whose lives revolve around a sense of location, seems to support this view. After all, the individual's familiar area map at birth is completely blank and is sketched out during life as the individual explores. If this is the view eventually we adopt, it will be the absolute antithesis of the old view of migration. It will also run contrary to the centre-of-party line amongst behavioural ecologists. According to this line, some ways of sketching the details of the familiar area map onto the blank page will be better than others. In which case, natural selection will operate to favour those individuals that happen to use the best rules for sketching out their maps. By now, evolution should have produced individuals that are innately predisposed to sketch out their map in a particular way. The map may be blank at birth, but the form of the map when complete will be the result of a lifetime's interplay between innate predisposition to sketch it in a particular way and, of course the features that the individual encounters.

At this point we encounter attack from the far left of the Behavioural

Ecology party. We become branded as pan-selectionist; as uncritically seeking an adaptive explanation for all features of animal biology. The voice of the far left (e.g. Lewontin 1977, Gould 1978) is of especial appeal to the proponents of human uniqueness. If not everything is adaptive, there do not have to be innate predispositions for all behaviour, and human culture in particular can be viewed as an escape from such adaptive predisposition. It then follows that, or so the argument goes, as most human behaviour is cultural, and the evolution of human behaviour is cultural evolution, human behaviour is freed from the constraints of natural selection. There is even a final irony to this argument, one that should not please the far left, though it is they that propose it: freedom from the constraints of natural selection could be adaptive! As cultural evolution is so much faster than genetic evolution, it is advantageous to be free of innate predispositions. These hinder any education that runs counter to them and can only be abolished by natural selection and genetic evolution. The fact remains, however, at the risk of being condemned as pan-selectionist, that if an innate predisposition encourages or speeds up the adoption of a learning process or pattern of behaviour that is advantageous, then it will evolve (given the genetic raw material on which selection can act). Everything depends on whether the advantages of this increased speed of learning outweighs any disadvantage that may result from hindrance in being able to learn something different. Hence, a pigeon may be innately predisposed to learn that the Sun moves across the sky; rarely will the pigeon's ancestors have suffered through finding it difficult to learn that the Sun stands still. Perhaps we should not dwell on the apparent predisposition of children in this same respect; or indeed the whole history of Man's understanding of the Solar System.

The critics of so-called pan-selectionism are trying to create a controversy where none exists. A few mainstream behavioural ecologists, perhaps, may claim that all behaviour is ultimately the product of genetic evolution but most would not. What they do claim is that pan-selectionism is a viable working hypothesis; the paradigm to be accepted in all cases except where proved invalid. This is neither ideology nor dogma but pragmatism; more than that, it is the scientific method. We have been through this argument before when, in Chapter 9, it was advocated that the new paradigm for migration should be that all animals have a sense of location and all that goes with it; that this should be assumed to be true in all cases except where proved to the contrary. Pan-selectionism and the exploration model are both working hypotheses and as such are comparable. I am as prepared to accept that there exists some behaviour that is not the result of genetic evolution as I am that somewhere there is a vertebrate that does not organise its life around exploration and a cerebral sense of location. First, however, I shall need proof that this is so, whereas I no longer demand proof that a particular behaviour is the product of natural selection or that a particular vertebrate explores and has a cerebral sense of location.

The opponents of pan-selectionism claim that the demonstration of adaptiveness for a particular characteristic of an animal rests more on the ingenuity of the investigator than it does on fact. Although often true, this is no argument against the 'adaptionist programme'. There is enough evidence in favour of genetic evolution and the adaptiveness of animal characteristics for pan-selectionism to be accepted as our paradigm. As far as behaviour is concerned, the outstanding demonstration of adaptiveness is the now-classic work of G.A. Parker (see brief review and references in Parker 1978) on the reproductive strategy of the Yellow Dung Fly, *Scatophaga stercoraria*. We are forced, as scientists, to accept the genetic adaptiveness of behaviour as our paradigm—not least because there is not a single piece of critical evidence to the contrary, despite the ingenuity of the far left.

Fig. 14.1 Male dung-fly, *Scatophaga stercoraria* resisting attempted take-over while guarding ovipositing female. (Photo by G. A. Parker)

We have no choice, therefore, but to do what we have failed to do so far and look for the innate, genetic predispositions that underlie migration behaviour. This we must do whether the life of the animal under consideration revolves round a sense of direction or location or whether the animal is an automaton. This means we must do so for Man as for all other animals. Perhaps by doing so we may at last see qualitative differences begin to emerge so that we may draw a line, either between Man and other animals or at least between vertebrates and invertebrates. Life is easiest, however, if we begin with the automaton for which we searched in the last chapter.

If an automaton exists among animals, its lifetime track will be, by definition, entirely the product of interplay between its innate predispositions and the immediate environment. At the risk of being unfair to barnacle larvae, let us assume, to give us an image about which to talk, that this organism is an automaton. We can begin with the release of the larva from its parent and the adoption of a planktonic way of life. The first thing it must do is attain a position in the water that is good for feeding. How does an automaton do this? Presumably selection has produced, mainly within the barnacle central nervous system, an innate 'image' of the sensory input the young larva should receive. If the larva can match its sensory input to this innate mental image, this templet (relating, perhaps, to hydrostatic pressure, light intensity, water taste/smell, water coloration, and so on), then it will be in the best position for feeding. Also encoded within the individual's genetic material will be instructions concerning how to home in on this innate image, e.g. if hydrostatic pressure is too great, migrate upwards; if light intensity is too great, migrate down, and so on. This in turn means that the animal must have an innate appreciation of up and down, or does it? It could achieve the same end by learning during the first brief phase of life that when it shifts in one direction relative to gravity light intensity increases, in the opposite direction it decreases, and at right angles it does not change. However, an innate predisposition to migrate up when light intensity is too low saves having to learn that this is what to do, or at least speeds the learning process. However, as we are assuming our barnacle larva to be an automaton, we should expect it to have innate instructions on how to correct differences between sensory input and innate templet, rather than having to learn how to do so.

If the best place in the water to occupy changes with time of day we should expect natural selection to have favoured mechanisms whereby the innate image of preferred sensory input also changes. In all probability, change in image with time of day will be under the control of an innate programme that works on a 24-hour cycle. The programme will need phasing with real time, of course, and we might expect this to occur during the first few days of larval life (perhaps using the light/dark interphase as the phase-setter) or perhaps even while in the parent (using daily fluctuation in the Earth's magnetic field?).

The best position in the water may also vary according to stage in the cycle of collection and digestion of food. According to the various feeding models for vertical migration cycles by zooplankton (e.g. Pearre 1973; Baker 1978a, Chapter 20), when the animal is hungry the best position is higher in the water than when satiated. We should expect, therefore, some influence of amount of food in the gut and/or stage of digestion on the mental image of preferred sensory input, this influence being superimposed on any variation in image that is linked to the animal's 24-hour clock.

If the best place in the water to occupy changes with age, we should again expect natural selection to have produced individuals with an innate programme for change in the preferred sensory input that is linked either to time (number of 24-hour cycles?) or to stage of development. If the best place to be changes with time of year, preferred sensory input should be linked to an appreciation of photoperiod and/or temperature. This particular sensory input may not only influence preferred sensory input directly but may also modify the daily and ontogenetic programmes.

Our barnacle larva, therefore, consists of a series of innate programmes, all whirring away within the animal, interacting with each other, being phased and/or modified by sensory inputs concerning time of day or year, and by feedback from internal organs such as the gut. Continuously they generate a preferred sensory input along with instructions on how to attain that input if the actual sensations coming from the outside world do not match. Our larva can be conceived as being involved in a continuous search to match actual and innately preferred sensory inputs: a true automaton.

When the stage of development and the times of year and day best for settling and metamorphosis approach, when the time is ripe for beginning the process of being a real barnacle, the programme for preferred sensory input—the preferred mental image—shows a major change. We should expect the animal to adopt the position that most increases its chance of being washed to shore to the position on the inter-tidal zone to which its species is adapted. We can imagine the innate programme for meta-morphosis being wound up and cocked, all ready to be set in motion by the simple trigger of the individual being able to match its new sensory preferences, (i.e. a hard substratum, surrounded by adult conspecifics) to an actual situation. Many will never receive that trigger, being washed up on a sandy instead of a rocky shore, or not being washed to shore at all. For the lucky few that do land on a rocky shore, the search to match innate preferences to actual sensations begins in earnest.

We can imagine that at first our larva has very stringent sensory preferences, those that would match the best possible site of all in which to live as an adult. Each of these preferences we can call a *habitat variable* because it is likely to relate in some way, either directly or indirectly, to habitat suitability. At intervals, the larva tests a site on the surface of the rock to see if it matches its innate image. If it does, it settles; if not, it moves on. Perhaps, if the barnacle larva really is an automaton, the search path used is also innately determined (see review of best search paths under different circumstances by Krebs 1978), sequences of right and left turns and responses to the habitat edge following each other in a way that produces the best possible search path over the rock face.

Suppose that the larva fails to match its stringent innate image to a real site; what happens? If it continues to search for the most perfect of all sites, it runs the risk of never settling, of never receiving the trigger to metamorphose. Clearly, the best strategy is for the innate image gradually to become less stringent as the larva searches, so that as time goes on less and less suitable sites match that image. We could put this another way and say that the animal's threshold for settling gradually decreases (or alternatively its *migration threshold* for continued search gradually increases); but threshold to what? Obviously, the threshold with respect to those habitat variables that contribute to the individual's set of sensory preferences. Perhaps at first the larva would accept as a match only those sites with fairly precise characteristics: traces of past or present occupation by conspecifics; conspecifics at a particular density, not too great, not too sparse; a rock face at a particular angle and with particular surface texture; rock of a precise chemical composition; and so on. The threshold is not a simple one, but *composite*, receiving contributions from many different habitat variables.

As time goes on the threshold changes. The innate preferences become less stringent, the tolerance levels to, say, the rock angle and/or density of conspecifics, become wider. Eventually, the larva encounters a site that matches its ever-widening image of a suitable site, the threshold for settling is exceeded, and metamorphosis is triggered. Our barnacle has played out its entire larval life by following innate instructions that allow it to find a position in the environment which matches its present, innately determined, sensory preferences. It responds only to the moment with no reference to its past: a true automaton.

We have said that the threshold for settling begins high, and then falls. We can say more about this threshold and the level at which it is set by natural selection. During evolution, the ancestors of our barnacle larva will have had a variety of such thresholds. Each time a barnacle settled in a site when it would have done better had it continued to search, it would leave fewer offspring than it could have done. Equally, each time a barnacle rejects a site but then fails to find a better one, it too leaves fewer offspring than it could have done. Obviously, no larva can predict whether it will find a better habitat by continuing to search. Evolution can work only on the mean expectation of further search or migration. Over the generations the individuals that do best are those that have a threshold that just matches the mean expectation of migration. This is the position at which we should expect natural selection to fix the migration threshold (see formal discussion in Baker 1978a, Chapter 8). Moreover, before being attacked by the critics of pan-selectionism, we can even present evidence that it has been fixed at this position (Baker 1978a, pp 226–7), again taken from Parker's excellent work on Dung Flies.

We seem to have been fairly successful in providing a viable automaton, an individual that can live its life entirely by responding to the way present sensory input matches, or fails to match, its present mental templet of how a good habitat should be, taking no account of the past. However, we have to be careful. Suppose our automaton has a migration threshold that is influenced by personal experience. Suppose that, as it searches for a suitable site in which to settle, it encounters a succession of very poor sites. This could mean that the mean expectation of migration in this place and time is lower than usual, and perhaps such experience modifies the migration threshold accordingly, speeding up the increasing tolerance in its mental image to allow for the poor prospects in this particular area. Cowie (1976) has suggested that in such a situation there should be a sliding memory 'window' whereby only the last few sites encountered have an influence.

Such continuous adjustment of the migration threshold, the tolerances in the mental templet, would be an undoubted effect of personal experience. However, the set of rules by which the mental templet would respond to the window of recent experiences would presumably also be innately programmed as would the size of the window itself. In other words, even the influence of past experience on the animal's mental image is dependent on innate instructions. We still have an automaton, but now it is an automaton influenced by previous experience: an automaton with a past.

Consider next an aphid. Where are the differences from the barnacle larva in the nature of interaction between innate programmes and the environment? Whether the aphid developed wings in the first place depended on the conditions experienced by its mother or itself and whether these exceeded the innate threshold for wing development. Whether or not it migrates, after developing wings, again depends on conditions on the host plant at the time the aphid becomes adult, or just before, and whether these exceed its innately fixed migration threshold. We can imagine that, when in the air, the aphid's flight behaviour reflects a continuous attempt to match its aerial environment to innately determined preferred sensations; that it is wound up to respond to perception of a habitat that matches its innate image of a suitable place to settle, responding to such a match by flying down into that area and beginning to search. All this can be based on innate predisposition and programming, the products of generations of natural selection.

The only major difference between the migration behaviour of an aphid and that of a butterfly, moth or grasshopper is that while they are migrating across country the latter three animals, as part of their image of preferred

Fig. 14.2 Wing polymorphism in the Rose Aphid, *Macrosiphum rosae*. (Photo by R. R. Baker)

sensory input, align their body relative to some celestial or geomagnetic orientation cue.

The arrangement of orientation cues into a hierarchy of preferences is presumably also innate. A moth will have an innate preference to maintain the Moon's disc at a certain angle relative to its body. Illumination of a particular set of ommatidia when the body angle is adjusted to be at right angles to gravity is part of the innate sensory input that the moth tries to match with its environment. Moreover, the light must be moonlight. Evidence is growing that moths respond to an artificial light source, as on a light trap, because they mistake the artificial light for the Moon (Baker and Sadovy 1978). It is not, however, a simple, automatic response to the only bright light source in the environment. The nearer the artificial light source mimics the real Moon in terms of elevation, apparent vertical diameter, and perhaps brightness and spectral composition, the more likely the moths are to respond. If no ommatidia are illuminated by moonlight, then the preferred innate sensory input switches to a particular pattern of small points of light: in other words, stars. If small points of light are also not available, then the preferred sensation switches to a particular align- ment of the geomagnetic sense organ. Along with these innate sensory preferences (i.e. preferred sensations; dare one say feelings?) we should expect innate instructions on what to do when actual sensations differ from the innate templet: what to do if flying too high, or too low; what to do when the wind is a head wind or a side or tail wind. All this could be innate, but is it?

Certainly, we should expect the butterfly or moth, etc. to be innately predisposed to respond in particular ways to the wind. Hence, temperate butterflies, such as the Small White, drop down to a lower height as wind speeds increase, no matter what the direction, whereas autumn Monarchs and arid-region tropical butterflies rise up to take advantage of tail winds. Does the individual also, however, finely tune its responses on the basis of personal experiences? Does it learn how low it should drop to compensate for a certain wind speed and direction? Does it learn that if it flies in the shelter of hedgerows it can maintain its preferred compass direction for less effort when the wind is strong than if it flies across open fields? Does a butterfly learn that if it flies too high with the wind it gets carried past a suitable habitat before it can drop down and so has to battle back against the wind? Is it predisposed to do all these things but has to learn the finer behavioural details to carry them out? Is the only difference between the butterfly and aphid, apart from the presence of an innate preferred compass direction, that the butterfly is predisposed to learn to respond to the wind and height above ground in one way whereas the aphid is predisposed to learn to respond in a different way?

Response to the wind may be one area in which the behaviour is not so much programmed as that the individual is predisposed to learn to behave in a certain way; response to stars and compensation for the Sun's movement across the sky are certainly others. We do not know the precise details of star orientation by the Large Yellow Underwing Moth, largely because moths caught in the field and brought directly into a planetarium for testing failed to accept the artificial sky as a substitute for the real night sky. As is standard practice for planetarium experiments on birds, the moths to be tested probably need to be reared under the planetarium sky in order to take the study further. If we accept the reality of star orientation by moths (Sotthibandhu and Baker 1979), failure to accept an artifical sky suggests that fairly subtle learning is taking place rather than simple orientation to, say, the brightest point in the sky. The most likely explanation, one that will do to make the point, is that each individual moth learns to recognise a particular pattern of stars. Undoubtedly, moths are innately programmed to observe and discriminate between star configurations, but the actual pattern that they use by which to orientate has to be learned.

If any butterflies, like bees, compensate for the movement of the Sun, they will undoubtedly be predisposed to learn that the Sun moves across the sky. The fine details of this movement, however, will have to be learned.

This may seem like a simple catalogue of learning by insect migrants. It may seem that we have simply switched from discussing pure instinct to discussing learning. What we have done, however, is more subtle than that. The point that is being made so tortuously is that, whether butterfly or aphid, insect or barnacle larva, the individual is a sentient creature. It has within itself, almost certainly within its central nervous system, no matter how small, an image of how things ought to *feel*. Put this in mechanistic terms of matching sensory inputs to innate templates if you wish. The fact remains that the individual is forever aware of how things should feel and spends its life chasing a position in its environment in which everything will feel as it should. In chasing these positions, the animal uses mechanisms that it was predisposed to use at birth and which may or may not be refined by personal experience. Such refinement would almost always seem to lead to greater efficiency than behaviour that is irrevocably fixed at birth.

It is a moot point whether the mental image of how the habitat for which these insects are searching should feel is 'conscious' or 'unconscious', and the distinction seems relatively unimportant. The important point is that the image exists, that the insect can recognise when the image is not matched by the environment, and that it can respond in a way that will improve the match. Most important of all, it can recognise when it has arrived at a suitable site. Moreover, the comparison between preferred and actual sensory input almost certainly takes place within the central nervous system. Given all this, is the image conscious or unconscious? Whatever the answer to this question, it is difficult, even for insects, to do other than accept their sentience and that their behaviour is the result of interplay between innate programming, innate predisposition, and refinements of both of these as a result of personal experience. How much does the situation change when we come on to consider animals with a sense of location?

The key behaviour in the attainment of a sense of location is exploration: the process of exploratory migration, habitat assessment, and navigation. At one level, exploration is no different from the behaviour already considered for animals not using a sense of location. If the best of an animal's feeding sites, say, do not match up to its innate mental image of how such a site should be then it begins to search for such a site. The only difference is that in searching for such a site, an animal with a linear range is not predisposed to return to its starting point, whereas an animal with a familiar area is so predisposed—until, that is, it finds somewhere which more nearly matches its innate image. Apart from this difference in predisposition to return, there is no fundamental distinction between exploration and NCR migration; in both the animal is seeking a habitat that matches an innate mental image. We have already seen how personal experience can influence at least the tolerances of the mental templet that an NCR migrant has of the site for which it is searching. The same personal experiences, whether it encounters a succession of good or poor sites, could influence the mental image of an explorer. Perhaps the only difference between an NCR migrant and an explorer is in the number of images. An NCR migrant needs only two mental images, one of how things ought to feel at a suitable site and one of how they should feel while searching. An explorer also has these two images. In addition, it carries around with it a mental image of the feeding site already being used. New sites encountered can then be judged against both of the site-related images.

If we multiply the number of images carried around by an animal, we encounter a qualitative leap to another feature associated with exploration but not with NCR migration: the collection of information for future use. It is easy to see how this evolved. Often something unexpected, beyond its previous experience, may happen to an animal, that suddenly necessitates moving elsewhere or making use of a different type of habitat from that used previously. Animals which have collected an amount of previously useless information during their travels should be able to counter such situations better than those which have not. They have more alternative places to which they can migrate. Of course, an animal can have too much useless information. There will be an optimum compromise between not enough and too much.

How can an animal be motivated to collect information that is not immediately useful? It must have a drive to explore and collect information simply for the sake of it. We should expect, therefore, some individuals of some species at some times to be innately predisposed to explore, solely in order to collect information; to build a familiar area of a size large enough to cope with future contingencies but not so large as to represent wasted effort.

Reaching the age of first reproduction is something of an emergency. By the time that age is reached, the animal needs ideally to have a home range suitable for reproduction. Before that age its immediate requirements have been different, usually less stringent. It follows that, again ideally, immaturity should be a phase of collecting information which although useless at the time will be useful in the future, often a long time in the future, when adult. The individuals that do best are likely to be those that spend their youth doing just that: collecting information for the sake of collecting information, not for its immediate usefulness. Translated into other terms, we should expect immaturity in animals with a sense of location to be a time of *programmed restlessness*, of exploration for the sake of exploration—innately directed preparation for the future contingencies of adulthood and reproduction. The irony is not hard to see. The most cerebral of all animals, those with a permanent sense of location, require the preparation brought about by an innate programme of predisposition to explore if they are to achieve this sense by the appropriate age.

What evidence is there that the adolescent phase of exploration that is such a marked feature of animals with a familiar area is under the influence of an innate programme for restlessness: an inbuilt urge to travel and visit new places just for the sake of travelling and seeing new places, to find out what they are like? Unfortunately, techniques for tracking individuals are rarely adequate for exploratory migration either to be followed or to be distinguished from other movements. Even if they were, the environment could not be controlled sufficiently to decide whether frequency of exploration was programmed. Fortunately, a method that allows the possibility of programming to be studied does seem to be available: the monitoring of wheel-running by caged animals.

I suggested previously (Baker 1978a, p. 379) that wheel-running by captive mammals could be used as a specific monitor of their urge to perform exploratory migration. Since then, the whole field of wheel-running has been reviewed (Mather 1981). While accepting that wheel-running does reflect the urge to carry out exploratory migration, Mather has extended the concept to include any occasion that the animal requires some resource that is not attainable within its cage or enclosure. In displacement experiments, for example, caged but wild-caught Woodmice, *Apodemus sylvaticus*, and Bank Voles, *Clethrionomys glareolus*, wheel-run in the direction of home.

The evidence gathered together by Mather is now overwhelming. Small mammals run in a wheel as a substitute for exploratory migration; they do not do so for exercise, in response to boredom, or as a reflection of general activity. This being so, if conditions in the cage are kept constant, any ontogenetic variation in wheel running can be taken to indicate that ontogenetic variation in exploratory migration in the natural world is under the control of an innate and endogenous programme. A typical ontogenetic curve for wheel-running is shown in Fig. 14.3. The shape of

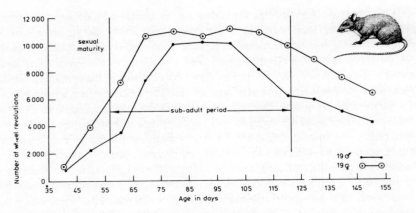

Fig. 14.3 Variation in wheel-running activity of Laboratory Rats in relation to age and maturity when maintained under constant conditions
Each dot shows a ten-day average for 19 males and 19 females. Wheel-running increases at the approach of sexual maturity, reaches a peak during sub-adulthood, and then declines at the normal age of first reproduction.
[Re-drawn from Richter (1933) with maturation data for wild Rats added from Calhoun (1963)]

this curve is so similar to the shape of the ontogenetic curve for exploratory migration deduced for a wide range of vertebrates (Baker 1978a, Chapter 17) that it can be taken to support the view that the adolescent phase of exploration in animals with a familiar area is under the control of endogenous programme. Selective breeding experiments with rats have demonstrated that this programme is genetically determined. Castration experiments have demonstrated at least some of the hormonal pathways by which the innate programme is mediated (reviewed by Mather 1981).

It seems that there is not only an *ontogenetic* programme for restlessness. Bank Voles, for example, show *seasonal* shifts in distribution that seem to be the result of seasonally triggered peaks of exploration. This is probably also the case for all animals that show seasonal peaks of exploration such as Woodmice, *Apodemus sylvaticus*, (Randolph 1977) and Deermice, *Peromyscus maniculatus*, (Fairbairn 1978). Bank Voles at least also show seasonal peaks of wheel-running. Superimposed on ontogenetic programmes for restlessness, therefore, may be seasonal programmes, perhaps triggered or phased by variation in daylength and temperature.

There is evidence, too that the most extreme cases of exploration, those performed by long-distances migrants, are also under the control of endogenous programmes for restlessness. The evidence, this time for birds, again has to be derived from studies of caged individuals.

It has long been known that when birds that are normally long-distance migrants are caged, they show increased activity at the same time of year that free-living conspecifics are migrating. This increased activity takes the form of hopping and fluttering restlessly around in their cages. *Migratory restlessness* (= *Zugunruhe*) is particularly conspicuous in normally day-active birds that migrate at night. The characteristics of migratory restlessness in birds are consistent with the view (Baker 1978a, Chapter 27) that it reflects exploratory migration threshold. On this view, performance of migratory restlessness indicates that the bird's exploratory migration threshold is low enough to be exceeded by its cage environment, whereas the absence of restlessness indicates that the threshold is high enough not to be exceeded by its environment. Cycles of restlessness in caged birds can be taken, therefore, to indicate cycles of variation in exploratory migration threshold—cycles of variation in the innately programmed urge to explore. Figure 14.4 shows that birds maintained in constant photoperiod over more than a year nevertheless continue to show circannual cycles of moult, fat deposition, and migratory restlessness. This indicates that, at least in broad terms, exploratory migration is under the influence of an endogenous programme. Fig. 14.4 also shows that, as expected, the amount of restlessness shown by birds in their first autumn and spring is greater than in later years.

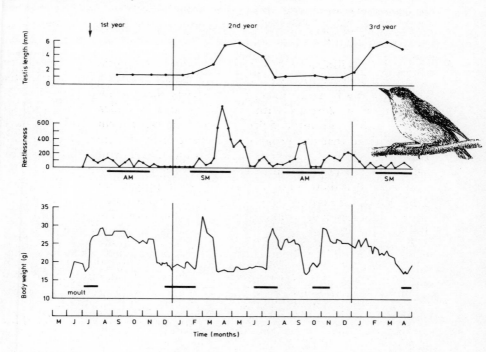

Fig. 14.4 Circannual rhythms of moulting, body weight, migratory restlessness, and testis length in a Garden Warbler, *Sylvia borin*
The bird was hand-raised before being transferred (arrow) to constant photoperiodic conditions of 10 hours light and 14 hours dark. Solid bars beneath the restlessness graph show the timing of migration (AM, autumn migration; SM, spring migration) for free-living conspecifics.
[Re-drawn and modified from Berthold (1978)]

Variation in amount of autumn migratory restlessness in Old World warblers of the genera *Sylvia* and *Phylloscopus* correlates well with the variation in migration distance between the same species. Furthermore, if maintained under the photoperiod conditions they would experience during migration in nature, the beginning and end of restlessness and the timing of the most intense periods of restlessness seem to coincide respectively with the beginning and end of autumn migration and the time during which free-living conspecifics cross the Mediterranean area and the Sahara Desert at maximum rate (Gwinner and Czeschlik 1978).

In many short-distance migrants the duration of autumn restlessness exceeds that of actual autumn migration by free-living conspecifics. Indeed, some demes of birds, such as the southern Pacific coastal race (*nuttalli*) of the White-Crowned Sparrow, *Zonotrichia leucophrys*, of North America, that show no standard autumn migration, nevertheless continue to show migratory restlessness. Migratory restlessness by caged birds that exceeds observed migration periods in nature would be expected if restlessness indicates that the bird's threshold has been exceeded by its cage environment. Such an environment is much more likely to exceed the exploratory migration threshold than is the natural environment. Hence, migratory restlessness is more likely to occur in a cage than is exploratory migration in the wild. It is interesting that given an appropriate photoperiod and, presumably, adequate food, water and temperature, migratory restlessness ceases for the winter period. Perhaps these conditions are sufficient to match the bird's mental image of what conditions should be in winter, in which case further exploration would not necessarily be initiated. In the natural world, a sudden onset of hard weather in winter will trigger further exploration.

We can pursue further this idea of an innate mental image of how a good habitat should feel. Experiments have demonstrated that Coal Tits, *Parus ater*, and Blue Tits, *P. caeruleus*, of Europe are innately predisposed, respectively, to prefer coniferous and deciduous trees for foraging (see review on habitat selection and recognition by Partridge 1978). There seems to be no direct evidence, however, that (as suggested when discussing barnacle larvae) a bird's innate mental image of what makes a good habitat varies with time of year. Nevertheless, it seems certain that this must occur. Throughout the autumn migration of a young bird, the major advantage lies in the collection of information: to locate, assess and rank a sufficient number of transient home ranges to allow for contingencies in future years. Such collection of information for the sake of it should be under the control of an innate programme for restlessness and continue for an appropriate period irrespective of the conditions the bird encounters. The same argument can be applied to spring migration. An intriguing observation, however, is that the migratory restlessness of caged birds in spring extends far beyond the spring migration season, continuing on into late summer (Gwinner and Czeschlik 1978). Evidence presented below suggests that the migratory restlessness in summer has different characteristics from that during normal spring migration. The interpretation (Baker 1978a) of restlessness as an indicator of exploration would expect that spring migratory restlessness reflects exploration for spring transient home ranges whereas summer restlessness reflects exploration for a suitable breeding home range. Whereas it was concluded earlier that a cage environment could well match a bird's mental image of how a good winter home range should be, it is less likely to match the mental image of how a good breeding home range should be. For a start, most such experimental birds are isolated, without access to a member of the opposite sex. If the cage environment is made more suitable as a breeding habitat, migratory restlessness decreases or terminates (Gwinner and Czeschlik 1978). Full view of a member of the opposite sex in an adjacent cage from early May to early June causes both male and female White-Throated Sparrows to cease migratory restlessness. Restoration of a visual partition between the cages results in its reappearance.

While free-living conspecifics are performing seasonal migrations, not only do caged birds show migratory restlessness, they also adopt a compass orientation appropriate to migration at that time of year. Experiments on the orientation of caged birds during restlessness have been reported by Perdeck and Clason (1974) and Gwinner and Wiltschko (1978). The birds concerned (Chaffinches, *Fringilla coelebs*, and Garden Warblers, *Sylvia borin*) showed a preferred compass direction that in the autumn and spring migration months were in the standard direction for the demes from which they were taken. In the Chaffinch, it seems that the direction preference is linked to changing day length. The compass cues used by birds have already been listed (Chapter 10). The changeover from a predisposition to migrate in the standard autumn direction to a predisposition for the standard spring direction seems in part to be pre-programmed and linked to the circannual biological clock and in part to be triggered by photoperiod and temperature. Not only preferred direction but also premigratory fat deposition, gonadal growth, and migratory restlessness are all finely phased by photoperiod and in some cases, temperature, during late winter to early spring, as well as being part of a circannual programme. During the extended period of spring/summer migratory restlessness, a consistent preferred direction is restricted to the normal period of spring migration. In summer, during the normal breeding season, directional preferences are inconsistent. This is as expected on the interpretation of migratory restlessness given above.

In Chapter 9, it was suggested that more birds than expected may establish a basic familiar area by association with, or observation of, conspecific adults, before beginning independent exploration. We may ask what happens if innate predisposition to migrate in a particular direction is ever different from the directions being flown by adults. Cross-fostering experiments on gulls suggest that innate predisposition was overridden by adult example. Juvenile White Storks, *Ciconia ciconia*, from East Germany have an innate preference to migrate south-east in autumn (Fig. 14.5). When displaced experimentally from East to West Germany, juveniles released before the local adults had departed from the area of release became part of the flocks of West German Storks migrating to the south-west. When the native Storks had departed further releases of East German juveniles produced recaptures to the south-east (Schüz 1951). Here is field support for the existence of an innate preferred compass direction for exploration as well as evidence that such a disposition is overridden by adult example.

We can now piece together a fairly clear story of the way that innate and learned factors interact during the first year of a bird's life while it is building up the familiar area and map within which most future migrations will take place. The young bird is born with an innate programme for growth and a crude circannual programme which will prime and phase moulting, fat deposition, changes in migration threshold (and, therefore, migratory restlessness) and changes in the mental image of how a good

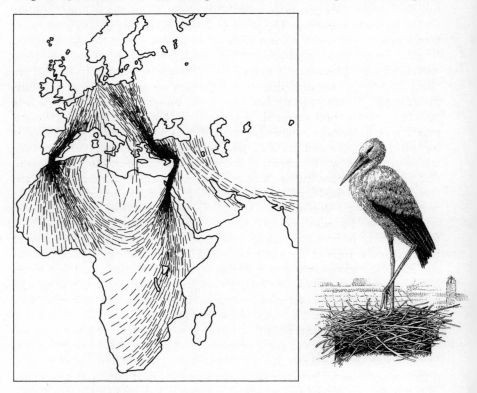

Fig. 14.5 The migration of the White Stork, *Ciconia ciconia*
[From Baker 1978a), after Verheyen (modified from Rüppell)]

habitat should feel. This crude circannual programme will be phased by environmental variables such as photoperiod and temperature. These will also trigger the innate programme that produces endogenous preferred compass directions for autumn and spring migrations. There is no difficulty in understanding how a bird can establish a basic familiar area by association with adults. When exploring independently, however, a young bird first collects information by flying along its preferred compass direction, then explores back along and sideways from its original track, collecting detailed information, before next continuing in its preferred direction. Caged birds, even without displacement, on some nights show a preference during restlessness for compass directions other than the standard direction (Rabøl 1978), thus suggesting that reverse and sideways exploration may also to some extent be programmed. Reaction to weather, particularly wind, should also be based on innate predispositions but perhaps will also improve with experience, the birds learning to become efficient weather forecasters. Exploration, pushing the familiar area further and further in the preferred autumn direction, continues until the endogenous programme switches to a mental image of how a good *winter* habitat should be. The bird then continues to explore in its preferred autumn direction until it encounters a habitat that, by virtue of day length, temperature, vegetation, food, water and other habitat variables, matches the bird's innate, mental image. Spring migration shows the same characteristics as autumn migration, though on a different part of the programme. When, towards late spring, the programme changes towards breeding, the mental image of a breeding habitat that the bird tries to match is no longer the innate one, though this image remains as a yardstick by which to judge real habitats. Instead, this image is superseded by the image of the best site for breeding found the previous autumn during post-fledging exploration. Instructions on how to match that mental image are stored on the learned familiar area map. Annual programmes of breeding, moulting and fat deposition, finely tuned to the individual's particular environment by photoperiod, temperature, success at breeding and feeding, etc. continue to produce a mental image of the best place to be at each particular time of day and year. Now, however, all images are those of the best places discovered during exploration in previous years. Only if this best place falls outside of the range of tolerance of the innate mental image is there an additional burst of exploration to see if there is not a better place available, a nearer match to the innate image.

How far have we travelled from our picture of the searching barnacle larva? Both bird and larva are forever striving to match their sensory input to the mental image they have of the best site available to them. The only difference is that the barnacle larva, we have assumed, uses a purely innate image whereas the bird adds to this image one based on personal experience (and perhaps on social communication). However, in order to obtain this new image, one based on familiarity, the bird behaves according to innate programmes that direct the timing of, and orientation during, exploration as well as variation in the innate image of the sensations for which to search.

Birds are the only long-distance migrants for which it has been shown that individuals have a preferred compass direction for exploration. Young Rainbow Trout, *Salmo gairdneri*, respond to experimentally produced long days by moving downstream and to short days by moving upstream (Northcote 1958). Salmon fry respond to the Earth's magnetic field by orientation in a direction that will take them to the traditional nursery lake for that part of their natal river system (Quinn 1980). If, as suggested in Chapter 8, this is part of exploration by young fish, it seems that preferred direction of exploration relative to the current or magnetic field is the result of an innate predisposition phased to the seasons by response to photoperiod or to stage of development.

As soon as young sea turtles hatch from the egg and dig their way out of the sand in which they were buried, they head directly to the sea and enter the water. Green Turtle, *Chelonia mydas*, hatchlings transferred to a beach in which the sea lies in a different compass direction from the sea at the home site nevertheless orient toward the sea (Carr and Ogren 1960). The migration is not, therefore, oriented in a preferred compass direction but must instead be based on some innate response to the beachscape. It now seems (Carr 1972, Mrosovsky 1978) that the hatchling orients directly toward the open, brighter horizon. Blindfold hatchlings fail to orient toward the sea and those blindfold over only one eye go round in circles. When they enter the water, whether by day or night, they orient 'away from land' (Ireland, Frick and Wingate 1978) and continue to travel 'away from land' in straight lines or gradual curves. When released from a beach facing another shoreline, the hatchlings reorient until once more travelling away from land. The orientation is maintained accurately, even after the land has disappeared beneath the visual horizon, thus suggesting that at some stage the 'away from land' direction is transferred to some other reference system, such as a celestial or geomagnetic compass.

Orientation toward an open, brighter horizon or away from a less-open, less-bright, horizon by the hatchling must be based on an innate predisposition. The sequence of intensive running and swimming immediately after hatching seems also to be based on an endogenous programme, for in an aquarium hatchlings continue to swim vigorously for 24 h after hatching (Carr 1972). Post-hatchling Green Turtles reared in captivity are carnivorous during early life, feeding on small marine invertebrates, before switching to a vegetarian diet with a preference for a few species of marine spermatophytic plants. Such a change in predisposition must surely be reflected by a change in the innate mental image of how a good habitat should feel.

Long-distance migrants such as whales, large ungulates, and human pastoral nomads may not have an innate preferred compass direction like birds. The young of these animals are probably never faced with the

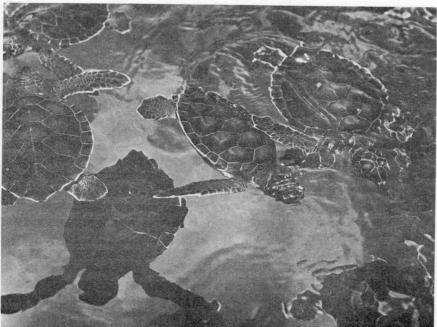

Fig. 14.6 The Green Turtle, *Chelonia mydas,* (Photos by Paul Turner)

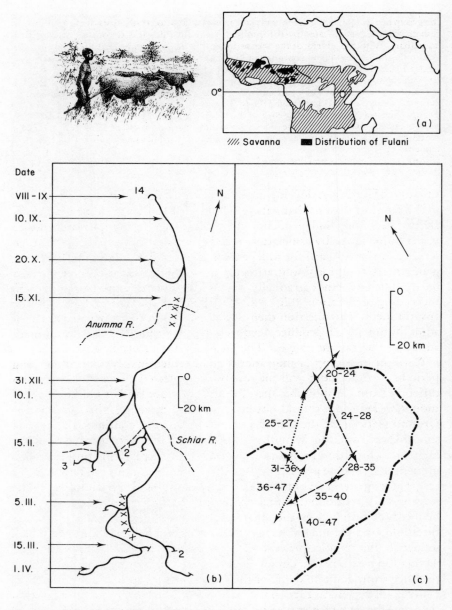

//// Savanna ■■ Distribution of Fulani

Fig. 14.7 Seasonal migration circuit of the Fulani and an example of shift in the migration circuit with time

(a) Distribution of Fulani pastoral nomads in relation to the African savanna

(b) Seasonal return migration of a group of 14 families, the males of which share common ancestry

At the peak of the wet season, the families are aggregated at the northernmost point of the circuit. At the beginning of the dry season, which starts earlier in the north than the south, the southward migration begins. Standing water persists along the route until January when

the group begins to disperse. Eventually, the group splits into family units though some may remain together as indicated by the numbers at the final dry season roosting and grazing areas (bar). With the advent of the wet season at the beginning of April (in the south) the return migration begins and continues until the groups are north of the main areas of risk to Tsetse Flies. During the dry season the areas of tsetse risk are localised (x).
(c) Shift in migration circuit over a period of nearly 50 years. Straight lines join the extreme wet and dry season limits of a particular migration circuit. Solid lines, migration circuit of original group; dotted and dashed lines, migration circuits after dichotomy into two groups. The heavy line marks the edge of the 1300 m high Jos Plateau. The sequence starts at year o (1880). Numbers by the lines indicate the period, in years after year o, that a particular migration circuit was used. A gradual southward drift of the migration circuit is evident.
[From Baker (1978a), after Stenning]

---

situation of having to develop independently a large familiar area that takes in the range of seasonal habitats to which their species or deme is adapted. Before becoming independent, such mammals are indoctrinated into a year's home range by association with parent(s) or other adults. This then serves as a springboard for independent exploratory migration. Over the generations, traditional migration circuits may thus remain unchanged or be modified whether gradually or rapidly by the experiences of each succeeding generation (Fig. 14.7c) depending on the stability of the environment. The question then is: does adoption of a particular form of adult home range produce feedback natural selection on the animals adopting these home ranges? The answer must surely be that it does.

Caribou that adopt a migration circuit which involves spending long periods on the tundra will inevitably be subjected to selective pressures different from Caribou that spend all year in woodland or on the slopes of mountain ranges. It should not be surprising, therefore, that the Barren-Ground Caribou of the Canadian tundra is usually considered a separate sub-species from the Woodland Caribou of the Canadian forests, the ranges of which are adjacent to each other. If the exploration programme is influenced by innate predispositions, these predispositions can be as much a target for genetic evolution as morphology and physiology. Quite possibly, as a traditional migration circuit becomes established, there may be feedback selection on the innate programmes of restlessness, migration threshold, and the mental images of how a good habitat should be which influence the whole process of exploration, habitat assessment, and navigation based on that circuit. These innate programmes may well come to differ from deme to deme of the species. How rapidly such differences evolve will depend in part on the rate of deme to deme removal migration (i.e. gene flow), which in turn will be influenced by the extent of ecological and genetic differences between the demes.

It is possible, therefore, that if a Barren Ground and Woodland Caribou were hand-raised from birth and then released away from herds of their natural demes, the Barren Ground Caribou would form a larger familiar area, perhaps oriented north-east/south-west, with a predilection to live on the tundra in summer. On the other hand, it may not. As both species

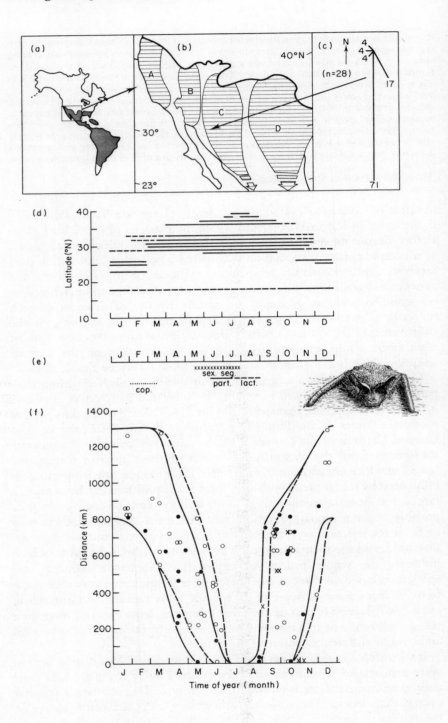

Fig. 14.8 Seasonal return migration of the American Free-Tailed Bat, *Tadarida brasiliensis*
(a) Distribution
(b) Four demes, two of which (A, B) migrate only short-distances between summer and winter sites and two of which (C, D) migrate long distances as in (c) to (e).
(c) Direction ratio of a group of individuals marked at summer roosts in Arizona. Length of line and number indicates the percentage of individuals that was recaptured in each of eight directions from the summer roosts at which the bats were marked.
(d) Latitudinal distribution with season of demes C and D as determined by observation. Solid line, common; dashed line, present but scarce.
(e) Seasonal behaviour: cop., copulation; part., parturition; lact., lactation;sex.seg., period of maximum segregation of the sexes with females occupying lower and warmer and males higher and cooler altitudes.
(f) Distance of recapture of individuals from demes C and D from the summer roost in which they were marked ● male; o, female; x, juvenile.
[From Baker (1978a), after various sources]

---

would normally form their basic familiar area by association with adults, it may be that selection has never had the opportunity of favouring different exploration programmes and mental images in the same way that it has morphology and physiology. The same programme would be quite adequate in both environments, given the role of social communication. The same would be true for different demes and species of humans, whales, and a few pinnipeds. For humans, would it be advantageous for the programmed restlessness of adolescent hunter−gatherers and pastoral nomads to be different, more pronounced, more seasonal than that of industrialists?

On the other hand, in mammals such as bats (Fig. 14.8) and various pinnipeds for which social communication may play little if any part in the establishment of a basic familiar area, the innate exploration programme as it relates to distance and direction may indeed differ from species to species and deme to deme. This may also be the case for amphibians, reptiles and fish. We should remember, however, that exploration programmes are programmes of predisposition. They are programmes of restlessness, preferred direction, and preferred timing. As such, they are flexible. What may be the normal course of exploration for a species or deme can be modified to an extreme. We cannot doubt, for example, that the programmed predispositions to explore of Herring and Lesser Black-Backed Gulls born in Britain are quite different (Chapter 9). Yet actual explorations can be interchanged by cross-fostering.

A different question is the way that exploration programmes may differ from individual to individual. The direction ratio of birds and butterflies is a case in point. Within a deme, different individuals have different preferred compass directions. It has been argued that the direction ratio evolves until it is in an evolutionarily stable state (Baker 1969, 1978a): the optimum compromise between all individuals flying in the best direction and the disadvantage that would result if they did so. At first sight, this may seem to be a group selection argument, but it is not. Initially, we might expect parents to pass on to their offspring a genetic predisposition to migrate in the same direction as themselves. In which case, the first equilibrium situation consists of lineages, each with a predisposition to migrate in a particular direction, in proportion to the optimum direction ratio. No single lineage can spread through the entire population. If it increases in numbers beyond a certain level it is selected against. Any lineage that is temporarily underrepresented is favoured by selection until it reaches equilibrium proportions once more. Once such an equilibrium situation has evolved, it can only be further invaded by a genetic lineage, each individual of which produces offspring that as a group have a direction ratio that matches the optimum. The existence of such a form of inheritance has been found in the Small White Butterfly (Fig. 14.9). The main point of interest for present purposes, however, is that this particular ESS (Evolutionary Stable Strategy: Maynard Smith 1974) implies that individual differences in innate programmes have evolved in such a way that all programmes are equally adaptive. No individual suffers a disadvantage solely as a result of having a particular preferred compass direction.

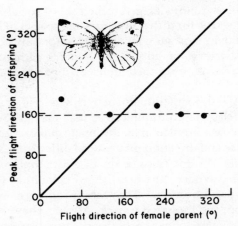

Fig. 14.9 The inheritance of mean migration angle and direction ratio in the Small White, *Pieris rapae*

Female Small Whites were captured during straight-line flight across a field, 3 for each of 8 migration directions relative to the Sun's azimuth. Virus infection killed the offspring of females for 3 of the 8 directions. When adult, the offspring were given 2 days at 17 hours daylight and 20 °C before being released in the middle of a large open field (different batches in different fields on different days). The graph shows the mean migration angle (●) for the different batches of offspring as a function of the migration direction of the female parent. All angles are measured clockwise from the Sun's azimuth. The solid line shows the relationship expected if the female parent passes on to her offspring a bias towards her own particular migration direction. The dashed line shows the relationship expected if each female produces offspring with a direction ratio the same as the population as a whole. Direction ratios relative to a mean angle of 159° varied for different batches from 58:0:33:9 to 43:21:21:15. None was significantly different from the direction ratio of 42:21:21:16 that characterises the free-living population.

[From Baker [1978a]]

The same argument can be applied to variation between individuals in the distances they are predisposed to migrate. Natural selection would be expected to produce a particular frequency distribution of migration distances: a distribution that represents the ESS for that particular species or deme. This distribution may be continuous, bi- or multi-modal, leptokurtic, and so on. The result is that some individuals are found to be more predisposed to migrate than others, more restless. Again, however, no one type of behaviour should be more advantageous than any other, all individuals doing equally well as long as the frequency distribution remains optimal. Longer-distance migrant individuals may expend more time and energy but reap that benefit of being surrounded by fewer competitors, perhaps in a better place.

Lack (1943, 1944), with particular reference to the bimodal distribution of migration distance in birds (i.e. *partial migration* Fig 14.10) that leads to

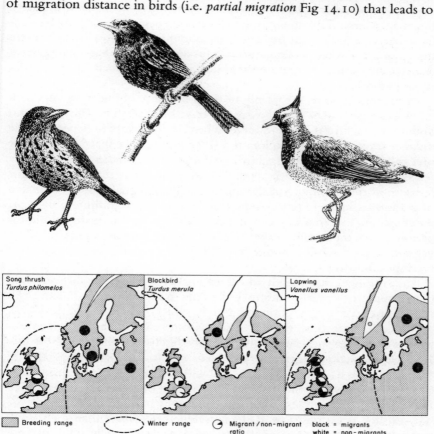

Fig. 14.10 Geographical and inter-specific differences in migrant—non-migrant ratio for three species of partial migrants
[From Baker (1978a), mainly after Lack]

some individuals being branded as 'migrants' and others as 'non-migrants' or 'residents', suggested there was an evolutionary balance between genes for migrants and non-migrants. In some years, migrants do better; in other years, non-migrants. Over a sufficiently long period of time, however, each type of behaviour does equally well. I prefer (Baker 1978a, pp. 635–6) to interpret partial migration as a frequency distribution in innate migration threshold; the range of tolerance in an individual's mental image of how a habitat should be. 'Migrants' thus have a lower migration threshold than 'residents'. They may also have a different mental image of what makes a good winter habitat (in the case of long–distance migrant birds). Evidence that 'migrants' are genetically different from 'residents' is available for small mammals (Myers and Krebs 1971) and birds (Greenwood *et al* 1979). As in all examples involving a threshold, whether or not an animal migrates depends on whether its habitat exceeds its threshold. Such a view of partial migration allows individuals to switch between being migrant and resident from one year to the next (in birds), behaviour that has now been found in many species. Parker has demonstrated for his dung flies that the frequency distribution of male migration thresholds is just that which gives all males equal pay-off.

Partial migration relates mainly to the probability of initiating migration in the first place rather than the distance travelled thereafter. The maximum distance that a bird is likely to travel when adult is likely to depend on the distance travelled during exploration when young. The speed and distance of exploration by a young bird, to judge from studies of migratory restlessness, seem to reflect endogenous programmes. Individual differences in the number and distance of 'legs' that make up the autumn migration are likely to reflect in part differences in programme and in part differences in preference over what combination of, for example, photoperiod and temperature makes a good winter habitat. Quite possibly, vagrants represent one extreme of this range of variation in programmed predisposition to explore.

The argument that all individuals do equally well, despite differences in innate migratory predispositions, holds only in a free situation (Chapter 8). It does not hold in a despotic situation in which individuals of low RHP (resource holding power) are excluded from resources by individuals of greater RHP. In such a situation it seems likely that individuals of low RHP will usually migrate further than individuals of high RHP (see formal analysis by Lomnicki 1978) and will also have a lower reproductive success. Nevertheless, the basic premise should be upheld that low RHP individuals do better by migrating or continuing to migrate than if they did not. In such situations, low RHP individuals do worse than high RHP individuals, not because of innate differences in migration programme, but because they are of low RHP.

There is one final way in which individual differences in migration programme could arise; one in which the individual may well find itself performing a migration that is disadvantageous to it. If so much of migration is based on innate predisposition, even for animals that live in a familiar area, the way is open for parents to manipulate their offspring to perform migrations that are disadvantageous to them. The mechanism of *parental manipulation* (Alexander 1974, Trivers 1974) has been discussed in the past mainly to cover the situation in which parents produce offspring with an optimal arrangement of altruistic tendencies so as to maximise the long-term reproductive success *of the parents*. Hamilton and May (1977) have shown that the theoretical optimum for the frequency distribution of migration thresholds among the offspring may be different for the parents and for the offspring. In such cases, the optimum for the parent is to produce a greater proportion of 'migrants' among the offspring than is optimum for the (migrant) offspring. Whether, in the real world, parents can enforce their optimum on their offspring or whether offspring can 'escape' parental constraint and produce their own optimum remains to be demonstrated.

One of the clearest examples of individual differences in innate migration programme involves the differences between males and females. Often these differences are such that one sex has on average a lower migration threshold than the other and is more inclined to leave one habitat for another and to travel further. All differences in migration threshold, whether inter- or intra-specific, inter-sexual, or ontogenetic, can be explained by a single model: the *initiation factor model*. This relates just three factors: migration cost; the probability that the habitat occupied is less suitable for the individual than a habitat elsewhere; and how much of the individual's total capacity for reproduction has already been achieved (Baker 1978a, Chapters 10 and 17). In the case of vertebrates, sexual differences in migration threshold are usually attributable to differences in migration cost and in the amount of 'investment' (time and energy, etc.) each sex makes in its particular home range (for a different view see Greenwood 1980). Other authors consider that a much more important factor is the risk of inbreeding.

When an individual mates with a closely related individual, the offspring are likely, on average, to be less fit than if the individual had mated with a less closely related individual. This 'inbreeding depression' results from the greater probability that the offspring will be homozygous for disadvantageous recessive genes. There is evidence for Great Tits, *Parus major*, that when siblings do pair, which happens rarely, they are less successful at fledging offspring (Greenwood *et al.* 1978). Similar data, albeit slight, is available for Baboons (Packer 1979). Greenwood and Harvey (1976) have argued for birds that if selection favours one sex staying near to its natal site to breed, this automatically imposes selection on the other sex to migrate a greater distance from the natal site before beginning to reproduce. A similar argument has been explored for humans (May 1979) while at the same time pointing out that, given sufficient incentive, humans are quite prepared to form incestuous relationships.

There are, of course, counter selective pressures. An individual that has survived to reproduce has some evidence that it is relatively well-adapted to the local environment. Offspring are more likely to be so adapted if the individual mates with an individual of similar phenotype. On this argument, immigrants should be less preferred as mates than individuals, particularly kin, that are also suited to the local environment. There is a further argument for mating with kin that derives from the inclusive fitness model of Hamilton (1964). The advantage of a given unit of parental care is related to the genetic similarity between offspring and parent, inevitably being more advantageous when directed toward a homozygous offspring than when directed toward a heterozygous offspring. Assortative mating (like mating preferentially with like) is common amongst animals and could reflect selection associated with these and other factors.

We may ask, then, as did E. O. Wilson (in May 1979), which is the most important factor as far as migration is concerned: inbreeding depression or assortative mating (of which mating with kin is an extreme form)? Whatever the answer we may also ask what I consider to be a more pertinent question: whether both are perhaps relatively unimportant, lifetime track patterns evolving in response to much more potent selective pressures?

In most birds, females have a lower migration threshold than males, travelling further, on average, from their natal site before settling to breed and being more inclined to change breeding sites from one year to the next. In others, such as ducks, the situation is reversed. Observed variation fits nicely with the relative investment made by the two sexes in their breeding home range (Baker 1978a, Chapter 17). In most birds, males establish and defend the breeding territory and the reproductive success of the pair is a function of the male's familiarity with that territory (Harvey *et al.* 1979). In ducks, males abandon the home range and the female after copulation, leaving care of the young to the female. If the risk of inbreeding is an important factor, female ducks should be much more loyal to their breeding home range in future years than females of other species; perhaps even as loyal as the males of other species. If relative investment is an important factor, female ducks should be little, if any, more loyal than the females of other species. A detailed comparison has not yet been made but, provisionally, data for location loyalty by female ducks (Sowls 1955, Shevaryova 1969) seem to support the importance of relative investment rather than the avoidance of inbreeding.

The evidence for primates is not clear either way. In many species, such as the Vervet Monkey, *Cercopithecus aethiops*, in which the males show a high rate of group to group migration, the females do not. In the Gorilla, *Gorilla gorilla*, most, if not all, females leave their natal group, which may be inherited later by one of their brothers (Harcourt, Stewart and Fossey 1976). On the face of it, this type of evidence may be used as an argument in

favour of avoidance of inbreeding. However, female Gorillas are reproductive on average 5 or 6 years before a brother is likely to inherit the
group. Moreover, any group the female joins must have a reasonable
probability of being taken over some time in the future by one of the
several brothers that become solitary. If the female group to group
migration is to avoid inbreeding at all, therefore, it is to avoid father–
daughter mating, not brother–sister. Yet I know of no evidence that
female Gorillas are more likely to leave a group still dominated by their
father than one in which the father has died and has been taken over by an
outside male. On the other hand, the decision to leave can be shown to be a
function of the male–female ratio in adjacent groups (Baker 1978a, pp.
92–94). When a multi-male group of Japanese Macaque, *Macaca fuscata*, is
provisioned and males no longer leave, there is no corresponding increase
in migration out of the group by females. Indeed, when a provisioned
group divides, the splinter-group, even if it is formed around an immigrant
solitary male, is more attractive to males than to females.

Whatever the reason for sexual differences in group to group migration
by primates, one thing is clear: females hardly ever, and in most species
never, perform solitary exploratory migrations over any distance whereas
males of all species do so. There is no direct evidence, but it seems likely that
this difference is the result of innate differences in migration programme
between the sexes.

The legacy from previous chapters with which we began this discussion
of learning and instinct in migration was strongly biased toward the role of
learning, particularly for animals with a sense of location. Now that we
have dissected the relationship, we find that beneath the surface film of
learning that is the sense of location there is a deep pool of innate
predisposition. The animals nearest to being automatons probably also
have a thin film of learning superimposed on a welter of innate
predispositions. The only difference between the two is that for the near-
automaton the film of learning is so thin that it is hardly noticed whereas for
the animal with a sense of location the film is so thick that it is only with
difficulty that the great pool of innate predisposition beneath it can be made
out.

So where does this leave Man? Is it credible that for Man alone there is no
deep pool of innate predisposition beneath the surface layer of learning?
Time and again in this book we have found similarities between the human
lifetime track and that of other animals. Now we find that even animals
with an apparently human sense of location gain this sense only as a result of
innate programmes of predispositions and instructions. It seems likely that
this will also be true for Man. What evidence can we present?

The answer is, of course, none. There are no absolutely critical data.
Despite improved techniques for analysing human behaviour in relation to
coefficients of relatedness (Eaves *et al* 1978), unequivocal proof that human
behaviour is influenced by inheritable, genetic predisposition is still

lacking. The design of the investigations concerned can always be criticised. Lacking, also, is proof that human behaviour is not so influenced. Nor can the study of migration provide *proof* of innate predisposition to behave in particular ways, only inferences.

Figure 14.11 shows the way that human migration incidence varies with age. The pattern shown is not a local one. Essentially the same curves are obtained for people the world over; they are a human characteristic. The similarity between this curve and the curve of wheel-running incidence in the rat is striking (Fig. 14.3), as is the similarity to the ontogenetic curves of migration incidence found in other mammals and other vertebrates. If we were dealing with any species other than Man we should conclude that our species had the same innate programme for restlessness during adolescence as these other species; an innate urge to travel and see new places. We should conclude that this programme was shared with other mammals;

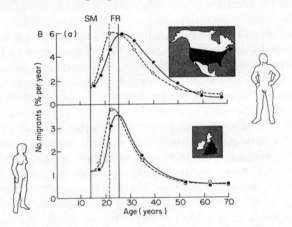

Fig. 14.11 Variation in migration incidence of humans in relation to age and maturity
The curves for the United States are for rural—urban migration and for England and Wales are for inter-regional migration.
Dashed curve, females; solid curve, males. SM, approximate age of sexual maturity; FR, average age of first reproduction (dashed line, females; solid line, males). Migration incidence increases after puberty, reaching a peak towards the end of sub-adulthood, and declining after the age of first reproduction.
[From Baker (1978a), after data in Thomas and the General Register Office]

indeed with other vertebrates; part of Man's vertebrate inheritance. However, as we are discussing Man, it seems we have to be more cautious, more scrupulous, and maintain that the case is 'not proven'.

As already mentioned, one of the characteristics of primates is that male adolescents are predisposed to go off on long, often solitary, explorations. Very few females behave in this way and if they leave their natal group at all they do so while it is in contact with a neighbouring group. Among human hunter—gatherers, pastoral nomads, and most agriculturalists, precisely the same pattern is found. Even among industrialists, females travel less far during adolescence and much less solitarily. Again, if it were not Man we were discussing we should conclude that such sexual differences in exploration behaviour were the result of innate predispositions, part of Man's primate inheritance. Again, as it is Man, we have to maintain the case is 'not proven'.

One of the enigmas of industrialisation is the massive flood of rural—urban migration that comes in its wake. All over the world, the advent of industrialisation, 'civilisation', and 'modernisation' has caused generation after generation of humans to stream away from the countryside into the towns and cities. The general improvement in the availability of food, the atmosphere of increasing optimism, and the multiplication of the population that comes with industrialisation, triggers people to migrate in their hordes to the densely populated cities; to migrate to premature death, reduced fertility, and impaired health — strange behaviour for Thinking Man: behaviour more lemming-like than that shown by Lemmings.

I have suggested previously (Baker 1978a, Chapter 14) that this enigma could be explained if it were assumed that humans, like other animals, have innate predispositions to use particular criteria to assess the suitability of habitats. In other words, humans also have innate mental images of how a good habitat should feel and during their adolescent explorations they search for locations that most nearly match that image. During most of human evolution, whenever food availability was good or improving, the best place to live was in the more densely populated areas; when bad or deteriorating, the best place to live was in the less densely populated areas. The innate mental image of good and bad habitats under the different conditions should have evolved accordingly. With the arrival of industrialisation, it was for many people the more-densely populated cities that most-nearly matched their innate mental image of a good site, even though that image was now at fault, no longer pointing the way to what was really the best place to settle: the countryside. Nevertheless, the people behaved according to the evidence of their innate mental image when deciding where to settle. Other people, in the local pockets of famine and oppression that also came with industrialisation, often joined the international migration stream, drawn to the images of wide open spaces associated with the United States and Australia during the nineteenth century. We are all aware that we carry within our minds mental images of places that we have visited and use these images in deciding where to go next. We also know that we have strong feelings about which places are good and which are bad. Moreover, our feelings about these places often differ from those of other people. Obviously, much of this response is based on past experience. How much is based on innate predisposition?

15

Epilogue

In this book, we have seen that many animals live their lives within a familiar area and have a cerebral sense of location comparable to that possessed by Man. Other animals, while not necessarily without a sense of location, do better by organising their lifetime tracks around a sense of direction. There may even be animals that live as 'automatons', responding only to the present. At times, these various ways of life seem very different; at others, the differences seem slight. Chapter 14 dissected these ways of life in an attempt to focus on differences in the interaction between learning and instinct. At the outset, it seemed possible that such dissection might highlight differences that were otherwise not apparent. In retrospect, the exercise seems to have done the opposite: it has made the different ways of life seem more similar.

All lifetime tracks can be conceived as the result of individuals moving through their environment in a continuous attempt to match their actual sensations to an ever-changing mental image of what those sensations should be. Individuals do this by following endogenous instructions on what to do when the two sets fail to match in a particular way. A true automaton would chase mental images that are engendered entirely within itself according to innate programmes and would use instructions that would also be entirely innate. In most, if not all, animals, however, mental images and instructions will be modified, either imperceptibly or grossly, by the individual's personal experiences, even though the rules for modification will probably also be innate. Such animals should retain their innate images as yardsticks by which to judge the modified images. Instructions and images that can be modified by personal experience are all that are necessary for life within a familiar area, given that the images are stored, not randomly, but in a spatial memory that reflects the spatial arrangement in nature of the sites to which they relate.

Viewed in this way, the evolution of the different ways of life is easy to conceive. No quantum leaps are required. All evolution of migration behaviour relates to slight modifications of innate mental templets, instructions on how to match these to the environment, rules on how to arrange the images and instructions within the central nervous system, programmes of variation in mental image *cum* migration threshold, and rules for learning on how to modify these in the light of personal experience. If such learning is as much under the control of innate instructions as the other aspects of behaviour, the convenient distinction used in this book between automatons and non-automatons breaks down.

Animals with a sense of location can be argued to be automatons as much as animals without: Man as much as a barnacle larva. At one level, therefore, all animals are automatons. At another level, all lifetime tracks can be reduced to the chasing of mental images.

Such reductionism facilitates an appreciation of the evolution of the different ways of life. We must beware, however, that the behavioural ecological view of migration does not suffer from a reductionism comparable to that which so plagued the age of ethology. Neither of the above levels, however, prevents qualitative differences from emerging at some higher level. All living organisms, for example, use similar chemicals, use similar rules for inheritance, and have similar basic metabolic processes. Yet we should wish to recognise a qualitative difference between an Elephant and a tree. In the same way, put programmes of mental images and instructions together in such a way as to produce an animal with a sense of location and we find behaviour qualitatively different from that of animals that do not use such a sense. Animals with a sense of location are no more sentient, no more cerebral, but they do behave differently. Moreover, we can only fully understand their behaviour if we study it at the level of an individual with a sense of location rather than at the level of an automaton chasing its own mental image.

All species are unique. Man is no exception. So where lies human uniqueness if it does not lie in any important sense in the mechanisms, even the cerebral mechanisms, involved in the lifetime track? If this book had been about reproductive behaviour, we should probably have concluded that there was nothing unique about human emotions. Books about genetic make-up, hormones, morphology, even language (Desmond 1979) may reach a similar conclusion. Only in one field can Man claim to be truly outstanding: the ability to manipulate and to manufacture objects. This outstanding ability is as much attributable to the possession of hands and an upright stance as it is to cerebral capacity, though presumably the three features evolved together. All other unique features: written language, long-distance communication, long-distance travel, navigational instruments, the human version of culture, the ability to change the environment so drastically, and many others, can be derived from this cerebral manipulative ability. Without it, Man would be just another species of primate, unremarkable in any way. We should expect, then, that any cerebral activity associated with manipulation and building may indeed be superior to that of other animals. That is no reason to expect superiority in any other feature, whether cerebral or emotional.

If we accept a gross similarity between Man and all other animals that live within a familiar area, we can use this similarity in our study of the lifetime track. Studies of other animals can tell us a great deal about the endogenous programmes and individual differences involved in human lifetime tracks. Equally, as in the study of navigation, experiments on humans may tell us far more about the migration behaviour of other animals than has ever previously been contemplated. Now that the study of migration has entered the age of behavioural ecology, it seems to have a future more promising and exciting, and more relevant, than at any other stage in its long history.

Glossary

Numbers indicate page on which a term is first defined or introduced

References and author index

The reference list in this book has been compiled with undergraduates in mind. Wherever possible, I have referred to the most recent review of the different fields rather than to the original works. Except where publications are of major or historical significance, I have also restricted reference to those that have appeared since the text of my last book was completed at the end of 1974. Where I have made use of data or information available before that date, I have not given reference on the assumption that a student wishing to follow up the information would refer back to my earlier review.

R. R. B.

The most important references for undergraduates are indicated by an asterisk (*). Numbers in brackets after each reference are the numbers of the pages on which the reference is quoted.

ALERSTAM, T. (1978) Analysis and a theory of visible bird migration. *Oikos* **30**, 273–349. (88, 90)

ALEXANDER, R. D. (1974) The evolution of social behaviour. *A. Rev. Ecol. Syst.* **5**, 325–383. (221)

AUBERT, M., AUBERT, J. and GAUTHIER, M. (1978) Telemediators in the marine environment. *Mar. Pollut. Bull.* **9**, 93–95. (117, 152)

BAKER, K. (1978) Vagrants in Britain. *Bird Study* **25**, 105–134. (90)

BAKER, R. R. (1968a) A possible method of evolution of the migratory habit in butterflies. *Phil. Trans. R. Soc.* B **253**, 309–341. (170, 172)

BAKER, R. R. (1968b) Sun orientation during migration in some British butterflies. *Proc. R. ent. Soc. Lond.* A **143**, 89–95. (172)

BAKER, R. R. (1969) The evolution of the migratory habit in butterflies. *J. Anim. Ecol.* **38**, 703–746. (172, 217)

*BAKER, R. R. (1978a) *The evolutionary ecology of animal migration.* Hodder & Stoughton, London. (8, 15–28, 32–46, 51–65, 68–71, 76, 79–81, 89, 92–95, 101–104, 108, 110, 113, 119, 121–2, 124–7, 133, 138, 141, 142, 147–155, 166–194, 198, 200, 205–225)

BAKER, R. R. (1978b) Demystifying vertebrate migration. *New Scientist* **80**, 526–528. (56, 90)

BAKER, R. R. (1980a) The significance of the Lesser Black-Backed Gull, *Larus fuscus*, to models of bird migration. *Bird Study* **27**, 41–50. (85–87, 107)

BAKER, R. R.(ed.) (1980b) *The mystery of migration.* McDonald/Harrow, London. (5, 193)

BAKER, R. R. (1980c) Goal orientation by blindfolded humans after long-distance displacement: possible involvement of a magnetic sense. *Science,* **210**, 555– 557. (131, 132)

*BAKER, R. R. (1981) *Human navigation and the sixth sense.* Hodder & Stoughton, London. (116, 124, 130, 131, 132, 134, 140)

BAKER. R. R., DRAPER, J., RAINEY, R. C. and WALOFF, Z. (1981) Comments at a meeting of the Royal Entomological Society of London. *Antenna,* **5**, 26– 27, 70– 72. (185)

BAKER, R. R. and MATHER, J. C. (1982) Magnetic compass sense in the Large Yellow Underwing Moth, *Noctua pronuba* L. *Animal Behaviour* (In press). (120, 173)

BAKER, R. R. and PARKER, G. A. (1979) The evolution of bird coloration. *Phil. Trans. R. Soc.* B **287**, 63– 130. (74)

BAKER, R. R. and SADOVY, Y. J. (1978) The distance and nature of the light-trap response of moths. *Nature, Lond.* **276**, 818– 821. (202)

BALDACCINI, N. E., BENVENUTI, S., FIASCHI, V., IOALÉ, P. and PAPI, F. (1978) Investigation of pigeon homing by means of 'deflector cages'. *In:* Schmidt-Koenig, K. and Keeton, W. T. (eds) *Animal migration, navigation and homing.* Springer, Heidelberg, pp. 78– 91. (129)

BARLOW, J. S. (1964) Inertial navigation as a basis for animal navigation. *J. Theoret. Biol.* **6**, 76– 117. (125)

BARNETT, S. A. (1971) *The human species: a biology of man.* (5th ed., revised) Harper and Row, London. (15)

BATSCHELET, E. (1965) *Statistical methods for the analysis of problems in animal orientation and certain biological rhythms.* American Institute Washington, D.C. (114)

BATSCHELET, E. (1978) Second-order statistical analysis of directions. *In:* Schmidt-Koenig, K. and Keeton, W. T. (eds) *Animal migration, navigation and homing.* Springer, Heidelberg. pp. 3– 24. (114)

BERGSTRÖM, U. (1967) Observations on Norwegian lemmings, *Lemmus lemmus* (L.) in the autumn of 1963 and spring of 1964. *Arkiv für Zoologi* **20**, 321– 363. (160, 161)

BERTHOLD, P. (1978) Concept of endogenous control of migration in warblers *In:* Schmidt-Koenig, K. and Keeton, W. T. (eds) *Animal migration, navigation and homing.* Springer, Heidelberg. pp. 275– 282. (207)

BOVET, J. (1978) Homing in wild myomorph rodents: current problems. *In:* Schmidt-Koenig, K. and Keeton, W. T. (eds) *Animal migration, navigation and homing.* Springer, Heidelberg. pp. 405– 412. (34, 136)

BROWNSTEIN, L. (1978) Sociobiology: the investigation of the biological aspects of social behavior and organization. *Sociology* **12**, 360– 368. (29)

CALHOUN, J. B. (1963) *The ecology and sociology of the Norway rat.* U. S. Dept. Health, Educ., Welfare. Public Health Service. (206)

CAMP, J. VAN and GLUCKIE, R., (1979) A record long-distance move by a wolf (*Canis lupus*). *J. Mammal.* **60**, 236–237. (157)

CARR, A. (1972) The case for long-range chemoreceptive piloting in *Chelonia. In*: Galler, S. R., Schmidt-Koenig, K., Jacobs, G. J. and Belleville, R. E. (eds.) *Animal orientation and navigation.* Scientific and Technical Information Office, National Aeronautics and Space Administration, Washington, D.C. pp. 469–483. (9, 212)

CARR, A. and COLEMAN, P. J. (1974) Seafloor spreading theory and the odyssey of the green turtle. *Nature, Lond.* **249**, 128–130. (9–11)

CARR, A. and OGREN, L. (1960) The ecology and migrations of sea turtles. 4. The green turtle in the Caribbean Sea. *Bull. Amer. Mus. Nat. His.* **121**, 1–48. (212)

CAVÉ, A. J., BOL, C. and SPEEK, G. (1974) Experiments on discrimination by the starling between geographical locations. *Progress Report 1973, Institute of Ecological Research, Royal Netherlands Academy of Arts and Sciences.* p. 82. (142)

CHARNOV, E. L. (1976) Optimal foraging: the marginal value theorem. *Theor. Popul. Biol.* **9**, 129–136. (168)

CHARNOV, E. L., ORIANS, G. H. and HYATT, K. (1976) The ecological implications of resource depression. *Amer. Natur.* **110**, 247–259. (170)

CHURCHILL, E. P. (1916) The learning of a maze by goldfish. *J. Anim. Behav.* **6**, 247–255. (60)

*CLUTTON-BROCK, T. H. and HARVEY, P. H. (1978) Group benefit or individual advantage? Introduction. *In*: Clutton-Brock, T. H. and Harvey, P. H. (eds) *Readings in Sociobiology.* Freeman, Reading. pp. 3–9. (73)

CODY, M. L. (1971) Finch flocks in the Mohave desert. *Theor. Pop. Biol.* **2**, 142–148. (170)

COWIE, R. J. and KREBS, J. R. in Krebs, J. R. (1978) (201)

CURRY-LINDAHL, K. (1962) The irruption of the Norway lemming in Sweden during 1960. *J. Mammal.* **3**, 171–184. (159)

DARWIN, C. (1873) Origin of certain instincts. *Nature, Lond.* **7**, 417–448. (125)

*DAVIES, N. B. (1978) Ecological questions about territorial behaviour. *In*: Krebs, J. R. and Davies, N. B. (eds) *Behavioural Ecology: an evolutionary approach.* Blackwell, London. pp. 317–350. (74, 79)

*DAWKINS, R. (1976) *The selfish gene.* Oxford University Press, Oxford. (73)

*DAWKINS, R. and KREBS, J. R. (1978) Animal signals: information or manipulation? *In*: Krebs, J. R. and Davies, N. B. (eds) *Behavioural Ecology: an evolutionary approach.* Blackwell, London. pp. 282–312. (73, 78)

DELIUS, J. D. and EMMERTON, J. (1978) Sensory mechanisms related to homing in pigeons. *In*: Schmidt-Koenig, K. and Keeton, W. T. (eds)

Animal migration, navigation and homing. Springer, Heidelberg. pp. 35 – 41. (134)

DESMOND, A. (1979) *The ape's reflexion.* Blond and Briggs, London. (29, 64, 70, 227)

DORST, J. (1962) *The migrations of birds.* Heinemann, London. (6)

DRAPER, J. (1980) The direction of desert locust migration. *J. anim. ecol.,* 49, 959 – 974. (185)

DWYER, P. D. (1966) The population pattern of *Miniopterus schreibersii* (Chiroptera) in north-eastern New South Wales. *Aust. J. Zool.* 14, 1073 – 1137. (55)

EAVES, L. J., LAST, K. A., YOUNG, P. A. and MARTIN, N. G. (1978) Model-fitting approaches to the analysis of human behaviour. *Heredity* 41, 249 – 320. (223)

EMLEN, S. T. (1972) The ontogenetic development of orientation capabilities. *In:* Galler, S. R., Schmidt-Koenig, K., Jacobs, G. J. and Belleville, R. E. (eds) *Animal orientation and navigation.* Scientific and Technical Information Office, National Aeronautics and Space Administration, Washington, D. C. pp. 191 – 210. (123)

*ENRIGHT, J. T. (1978) Migration and homing of marine invertebrates: a potpourri of strategies. *In:* Schmidt-Koenig, K. and Keeton, W. T. (eds) *Animal migration, navigation and homing.* Springer, Heidelberg. pp. 440 – 446. (122, 164, 192)

EVANS, P. R. (1972) Information on bird navigation obtained by British long-range radars. *In:* Galler, S. R., Schmidt-Koenig, K., Jacobs, G. J. and Belleville, R. E. (eds.) *Animal orientation and navigation.* Scientific and Technical Information Office, National Aeronautics and Space Administration, Washington, D.C. pp. 139 – 149. (150)

FAIRBAIRN, D. J. (1978) Dispersal of deer mice, *Peromyscus maniculatus.* Proximal causes and effects on fitness. *Oecologia* 32, 171 – 193. (206)

FERGUSON, D. E. (1971) The sensory basis of orientation in amphibians. *Ann. N. Y. Acad. Sci* 188, 30 – 36. (137)

FRAENKEL, G. S. and GUNN, D. L. (1940) *The orientation of animals, kineses, taxes and compass reactions.* Oxford University Press, Oxford. (7)

FRETWELL, S. D. (1972) *Populations in a seasonal environment.* Princeton University, New Jersey. (76)

FRISCH, K. VON (1967) *The dance language and orientation of bees.* Oxford University Press, London. (64, 122)

GAUTHREAUX, S. A. (1978) Importance of the daytime flights of nocturnal migrants: redetermined migration following displacement. *In:* Schmidt-Koenig, K. and Keeton, W. T. (eds) *Animal migration, navigation and homing.* Springer, Heidelberg. pp. 219 – 227. (150)

GEE, A. S., MILNER, N. J. and HEMSWORTH, R. J. (1978) The effect of density on mortality in juvenile atlantic salmon (*Salmo salar*). *J. Anim. Ecol.* 47, 497 – 505. (96)

GELPERIN, A. (1974) Olfactory basis of homing behavior in the Giant Garden Slug, *Limax maximus*. *Proc. Nat. Acad. Sci. USA* **71**, 966–970. (62)

GORZULA, S. J. (1978) An ecological study of Caiman, *Caiman crocodilus crocodilus*, inhabiting Savanna lagoons in the Venezuelan Guayana. *Oecologia* **35**, 21–34. (37, 69)

GOULD, J. L. (1975) Honey bee recruitment: the dance-language controversy. *Science* **189**, 685–693. (64)

*GOULD, J. L. (1980) The case for magnetic sensitivity in birds and bees (such as it is). *American Scientist* **68**, 256–267. (120, 133, 140)

GOULD, S. J. (1978) Sociobiology: the art of storytelling. *New Scientist* **80**, 530–533. (196)

GOULD, S. J. (1979) Nature seen: of turtles, vets, elephants and castles. *New Scientist* **81**, 100–102. (154)

GREENWOOD, P. J. (1980) Mating systems, philopatry and dispersal in birds and mammals. *Animal Behaviour*, **28**, 1140–1162. (221)

GREENWOOD, P. J. and HARVEY, P. H. (1976) The adaptive significance of variation in breeding area fidelity of the blackbird (*Turdus merula* L.) *J. Anim. Ecol.*, **45**, 887–898. (221)

GREENWOOD, P. J. HARVEY, P. H. and PERRINS, C. M. (1978) Inbreeding and dispersal in the great tit. *Nature, Lond.* **271**, 52–54. (221)

GREENWOOD, P. J., HARVEY, P. H. and PERRINS, C. M. (1979) The role of dispersal in the great tit (*Parus major*): the causes, consequences and heritability of natal dispersal. *J. Anim. Ecol.* **48**, 123–142. (220)

GRIFFIN, D. R. and BUCHLER, E. R. (1978) Echolocation of extended surfaces. *In:* Schmidt–Koenig, K. and Keeton, W. T. (eds) *Animal migration, navigation and homing*. Springer, Heidelberg. pp. 201–208. (117)

GRUBB, J. C. (1973) Olfactory orientation in *Bufo woodhousii fowleri*, *Pseudacris clarki* and *Pseudacris streckeri*. *Anim. Behav.* **21**, 726–732. (137)

GWINNER, E. and CZESCHLIK, D. (1978) On the significance of spring migratory restlessness in caged birds. *Oikos* **30**, 364–372. (208, 209)

GWINNER, E. and WILTSCHKO, W. (1978) Endogenously controlled changes in the migratory direction of the Garden Warbler (*Sylvia borin*). *J. Comp. Physiol.* A 125, 267–274. (209)

HAARTMAN, L. VON (1968) The evolution of resident versus migratory habit in birds: some considerations. *Ornis fenn.* **45**, 1–7. (86)

HAMILTON, W. D. (1963) The evolution of altruistic behaviour. *Amer. Natur.* **97**, 354–356. (73)

HAMILTON, W. D. (1964) The genetical theory of social behaviour. I, II. *J. theor. Biol.* **7**, 1–52. (222)

HAMILTON, W. D. and MAY, R. M. (1977) Dispersal in stable habitats. *Nature, Lond.* **269**, 578–581. (221)

HARCOURT, A. H, STEWART, K. S. and FOSSEY, D. (1976) Male emigration and female transfer in wild mountain gorilla. *Nature, Lond.* **263**, 226–267. (222)

*HARDEN JONES, F. R. (1968) *Fish Migration.* Arnold, London. (11, 98, 99, 101, 105, 106, 151)

HARRIS, M. P. (1970) Abnormal migration and hybridization of *Larus argentatus* and *L. fuscus* after inter-species fostering experiments. *Ibis* **112**, 488–498. (87)

HARTMAN, W. L. and RALEIGH, R. F. (1964) Tributary homing of sockeye salmon at Brooks and Karluk Lakes, Alaska. *J. Fish. Res. Bd Canada* **21**, 485–504. (98)

HARTWICK, R., KIEPENHEUER, J. and SCHMIDT-KOENIG, K. (1978). Further experiments on the olfactory hypothesis of pigeon homing. *In*: Schmidt-Koenig, K. and Keeton, W. T. (eds) *Animal migration, navigation and homing.* Springer, Heidelberg. pp. 107–118. (129)

HARVEY, P. H., GREENWOOD, P. J., PERRINS, C. M. and MARTIN, A. (1979) Breeding success of great tits in relation to age of male and female parent. *Ibis* **121**, 194–208. (222)

*HASLER, A. D. and SCHOLZ, A. I. (1978) Olfactory imprinting in Coho Salmon (*Oncorhynchus kisutch*). *In*: Schmidt-Koenig, K. and Keeton, W. T. (eds) *Animal migration, navigation and homing* Springer, Heidelberg. pp. 356–369. (151)

HASLER, A. D. and WISBY, W. J. (1951) Discrimination of stream odors by fishes and its relation to parent stream behavior. *Am. Nat.* **85**, 223–238. (9)

HAYNES, J. M., GRAY, R. H. and MONTGOMERY, J. C. (1978) Seasonal movements of white sturgeon (*Acipenser transmontanus*) in the mid-Colombia River. *Trans. Am. Fish. Soc.* **107**, 275–280. (37)

HEAPE, W. (1931) *Emigration, migration and nomadism.* Cambridge University Press, Cambridge. (v, 5)

HEINRICH, B. (1978) The economics of insect sociality. *In*: Krebs, J. R. and Davies, N. B. (eds) *Behavioural Ecology: an evolutionary approach.* Blackwell, London. 97–128. (64)

INGRAM, W. M. and ADOLPH, H. M. (1943) Habitat observations of *Ariolimax columbianus. The Nautilus* **56**, 96–97. (62)

IRELAND, L. C., FRICK, J. A. and WINGATE, D. B. (1978) Nighttime orientation of hatchling green turtles (*Chelonia mydas*) in open ocean. *In*: Schmidt-Koenig, K. and Keeton, W. T. (eds) *Animal migration, navigation and homing.* Springer, Heidelberg. pp. 420–429. (212)

JENKINS, D. (1978) Otter breeding and dispersion in Mid-Deeside, Aberdeenshire. *In*: Dunstone, N. (ed) Carnivore Biology and Management, Carnivore Group Symposium, Oxford University 28–29 Sept. 1978. Mammal Society, London. (53)

*JOHNSON, C. G. (1969) *Migration and dispersal of insects by flight.* Methuen, London. (6, 11, 13, 185, 187)

JONES, J. W. (1959) *The salmon.* Collins, London. (97)

KALLEBERG, H. (1958) Observations in a stream tank of territoriality and competition in juvenile salmon and trout (*Salmo salar* L. and *S. trutta* L.). *Rep. Inst. Freshwat. Res. Drottningholm* **39**, 55–98. (97)

KALMIJN, A. J. (1978) Experimental evidence of geomagnetic orientation in Elasmobranch fishes. *In*: Schmidt-Koenig, K. and Keeton, W. T. (eds) *Animal migration, navigation and homing*. Springer, Heidelberg. pp. 347– 353. (117, 120)

KANZ, J. E. (1977) The orientation of migrant and non-migrant butterflies, *Danaus plexippus* (L.). *Psyche* **84**, 120–141. (172, 175, 178)

KEETON, W. T. (1972a) Effects of magnets on pigeon homing. *In*: Galler, S. R., Schmidt-Koenig, K., Jacobs, G. J. and Belleville, R. E. (eds) *Animal orientation and navigation*. Scientific and Technical Information Office, National Aeronautics and Space Administration, Washington, D.C. pp. 579–94. (121)

KEETON, W. T. (1972b) Comments at a meeting. *In*: Galler, S. R., Schmidt-Koenig, K., Jacobs, G. J. and Belleville, R. E. (eds), *Animal orientation and navigation*. Scientific and Technical Information Office, National Aeronautics and Space Administration, Washington, D.C. p. 281. (121)

KEETON, W. T. (1979) Avian orientation and navigation: a brief overview. *British Birds* **72**, 451–470. (120)

KEETON, W. T., LARKIN, T. S. and WINDSOR, D. M. (1974) Normal fluctuations in the Earth's magnetic field influence pigeon orientation. *J. comp. Physiol.* **95**, 95–103. (135, 140)

*KELSALL, J. P. (1968) *The migratory Barren-ground Caribou of Canada*. Canadian Wildl. Serv. Monogr. Ser. 3, Canadian Wildl. Ser., Ottawa. (109)

KENYON, K. W. (1969) The sea otter, (*Enhydra lutris*) in the Eastern Pacific Ocean. *North Am. Fauna* **68**, 1–352. (158)

KIEPENHEUER, J. (1978) Inversion of the magnetic field during transport: its influence on the homing behavior of pigeons. *In*: Schmidt-Koenig, K. and Keeton, W. T. (eds) *Animal migration, navigation and homing*. Springer, Heidelberg. pp. 135–142. (129)

*KIRSCHVINK, J. L. and GOULD, J. L. (1981) Biogenic magnetite as a basis for magnetic field detection in animals. *Biosystems* **13**, 181–201. (120)

KLEEREKOPER, H., MATIS, J., GENSLER, P. and MAYNARD, P. (1974) Exploratory behaviour of goldfish, *Carassius auratus*. *Anim. Behav.* **22**, 124–132. (60)

KÖHLER, K.-L. (1978) Do pigeons use their eyes for navigation? A new technique! *In*: Schmidt-Koenig, K. and Keeton, W. T. (eds) *Animal migration, navigation and homing*. Springer, Heidelberg. pp. 57–64. (136)

KRADER, L. (1959) The ecology of nomadic pastoralism. *UNESCO Int. Social Sci. J.* **11**, 499–510. (15)

KRAMER, G. (1957) Experiments on bird orientation and their interpretation. *Ibis* **99**, 196–227. (141)

KRAMER, G. (1959) Recent experiments on bird orientation. *Ibis* **101**, 399–416. (141)

KREBS, J. R. (1977) The significance of song repertoires: the Beau Geste hypothesis. *Anim. Behav.* **25**, 475–478. (77, 83)

*KREBS, J. R. (1978) Optimal foraging: decision rules for predators. *In*: Krebs, J. R. and Davies, N. B. (eds) *Behavioural Ecology: an evolutionary approach*. Blackwell, London. pp. 23–63. (51, 106, 169, 199)

KREBS, J. R. and COWIE, R. J. (1976) Foraging strategies in birds. *Ardea* **64**, 98–116. (59)

KREITHEN, M. L. (1978) Sensory mechanisms for animal orientation–can any new ones be discovered? *In*: Schmidt-Koenig, K. and Keeton, W. T. (eds) *Animal migration, navigation and homing*. Springer, Heidelberg. pp. 25–34. (117, 119, 141)

KREITHEN, M. L. and KEETON, W. T. (1974) Detection of changes in atmospheric pressure by the homing pigeon, *Columba livia*. *J. Comp. Physiol.* **89**, 83–92. (117)

LACK, D. (1943) The problem of partial migration, *Br. Birds* **37**, 122–130. (219)

LACK, D. (1944) The problem of partial migration. *Br. Birds* **37**, 143–150. (219)

LACK, D. (1960) The influence of weather on passerine migration. *Auk* **77**, 171–209. (90)

LARKIN, S. and MCFARLAND, D. (1978) The cost of changing from one activity to another. *Anim. Behav.* **26**, 1237–1246. (168)

LARKIN, T. S. and KEETON, W. T. (1976) Bar magnets mask the effect of normal magnetic disturbances on pigeon orientation. *J. comp. Physiol.* **110**, 227–231. (140)

LEASK, M. J. M. (1977) A physico-chemical mechanism for magnetic field detection by migratory birds and homing pigeons. *Nature, Lond.* **267**, 144–146. (120)

LECREN, E. D. (1973) The population dynamics of young trout (*Salmo trutta*) in relation to density and territorial behaviour. *Rapports et Proces-Verbaux des Reunions Conseil International pour L'exploration de la Mer.* **164**, 241–246. (96)

LEE, P. B. and DEVORE, I. (eds) (1968) *Man the hunter*. Aldine, Chicago. (15)

LEFFLER, J. W., LEFFLER, L. T. and HALL, J. S. (1979) Effects of familiar area on the homing ability of the little brown bat, *Myotis lucifugus*. *J. Mammal.* **60**, 201–204. (136)

LEHNER, P. N. (1978) Coyote vocalizations: A lexicon and comparisons with other canids. *Anim. Behav.* **26**, 712–722. (70)

LEWONTIN, R. C. (1977) Caricature of Darwinism. *Nature*, **266**, 283–284. (196)

LINDAUER, M. (1955) Schwarmbienen auf Wohnungssuche. *Z. vergl. Physiol.* **37**, 263–324. (64)

LINDAUER, M. (1967) *Communication among social bees.* Harvard University Press, Cambridge, Massachusetts. (122)

LINDAUER, M, and MARTIN, H. (1972) Magnetic effect on dancing bees. *In*: Galler, S. R., Schmidt-Koenig, K., Jacobs, G. J. and Belleville, R. E. (eds), *Animal orientation and navigation.* Scientific and Technical Information Office, National Aeronautics and Space Administration, Washington, D.C. pp. 559–67. (120, 121)

LOMNICKI, A. (1978) Individual differences between animals and the natural regulation of their numbers. *J. Anim. Ecol.* **47**, 461–476. (220)

*MARSDEN, W. (1964) *The lemming year.* Chatto and Windus, London. (160)

MATHER, J. G. (1981) Wheel-running activity: a new interpretation. *Mammal Review.* **11**, 41–51. (205, 206)

MATHER, J. and BAKER, R. R. (1980) A demonstration of navigation by small rodents using an orientation cage. *Nature, Lond.* **284**, 259–262. (137)

MATHER, J. G. and BAKER, R. R. (1981) Magnetic sense of direction in woodmice for route-based navigation. *Nature, Lond.* **291**, 152–5 (120, 130, 135, 136)

MATTHEWS, G. V. T. (1955) *Bird navigation* (1st ed.). Cambridge University Press, London. (7, 114, 139)

*MATTHEWS, G. V. T. (1968) *Bird navigation* (2nd ed.). Cambridge University Press, London. (2, 106, 114, 120, 134, 139, 148)

MATTHEWS, G. V. T. (1973) Biological clocks and bird migration. *In*: Mills, J. N. (ed) *Biological Aspects of Circadian Rhythms.* Plenum, London. pp. 281–311. (122)

MATTHEWS, G. V. T. and COOK, W. A. (1977) The role of landscape features in the 'nonsense' orientation of mallards. *Anim. Behav.* **25**, 508–517. (114)

MAY, R. M. (1979) When to be incestuous. *Nature, Lond.* **279**, 192–194. (221, 222)

MAYNARD SMITH, J. (1964) Group selection and kin selection. *Nature, Lond.* **201**, 1145–1147. (73)

MAYNARD SMITH, J. (1974) The theory of games and the evolution of animal conflicts. *J. theor. Biol.* **47**, 209–221. (218)

MEAD, C. J. (1979) Colony fidelity and interchange in the sand martin. *Bird Study* **26**, 99–106. (56, 94)

MEAD, C. J. and HARRISON, J. D. (1979a) Sand martin movements within Britain and Ireland. *Bird Study* **26**, 73–86. (56, 57, 94)

MEAD, C. J. and HARRISON, J. D. (1979b) Overseas movements of British and Irish sand martins. *Bird Study* **26**, 87–98. (56, 94)

MERKEL, F. W. (1978) Angle sense in painted quails – a parameter of geodetic orientation? *In*: Schmidt-Koenig, K. and Keeton, W. T. (eds) *Animal migration, navigation and homing.* Springer, Heidelberg. pp. 269–274. (125)

*MILLS, D. (1971) *Salmon and trout: a resource, its ecology, conservation and management*. Oliver and Boyd, Edinburgh. (97)

MOREAU, R. E. (1972) *The Palaearctic-African bird migration systems.* Academic Press, London. (92)

MORRISON, D. W. (1978) Lunar phobia in a neotropical fruit bat, *Artibeus jamaicensis* (Chiroptera: Phyllostomidae). *Anim. Behav.* **26**, 852−855. (56)

MROSOVSKY, N. (1978) Orientation mechanisms of marine turtles. *In*: Schmidt-Koenig, K. and Keeton, W. T. (eds) *Animal migration, navigation and homing*. Springer, Heidelberg. pp. 413−419. (212)

MUELLER, H. C. (1968) The role of vision in vespertilionid bats. *Am. Midl. Nat.* **79**. 524−525. (136)

MYERS, J. H. and KREBS, C. J. (1971) Genetic, behavioral and reproductive attributes of dispersing field voles *Microtus pennsylvanicus* and *Microtus ochrogastor*. *Ecological Monographs* **41**, 53−78. (220)

MYERS, K. and POOLE, W. E. (1961) A study of the biology of the wild rabbit, *Oryctolagus cuniculus* (L.), in confined populations. II The effects of season and population increase on behaviour. *C.S.I.R.O. Wildl. Res.* **6**, 1−41. (52)

NISBET, I. C. T. and DRURY, W. H. (1968) Short-term effects of weather on bird migration: a field study using multivariate statistics. *Anim. Behav.* **16**, 496−530. (150)

NORTHCOTE, T. G. (1958) Effect of photoperiodism on response of juvenile trout to water currents. *Nature, Lond.* **181**, 1283−1284. (97, 212)

O'KEEFE, J. and NADEL, L. (1979) *The hippocampus as a cognitive map*. Oxford University Press, Oxford. (123)

OLTON, D. S. (1977) Spatial memory. *Sci. Amer.* **236**, 82−98. (123)

PACKER, C. (1979) Inter-troop transfer and inbreeding avoidance in *Papio anubis*. *Anim. Behav.* **27**, 1−36. (221)

PAPI, F., FIORE, L., FIASCHI, V. and BENVENUTI, S. (1972) Olfaction and homing in pigeons. *Monitore zool. ital. (N. S.)* **6**, 85−95. (126, 138)

*PAPI, F., IOALÉ, P., FIASCHI, V., BENVENUTI, S. and BALDACCINI, N. E. (1978) Pigeon homing: cues detected during the outward journey influence initial orientation. *In*: Schmidt-Koenig, K. and Keeton, W. T. (eds) *Animal migration, navigation and homing*. Springer, Heidelberg. pp. 65−77. (129)

PARKER, G. A. (1974a) Courtship persistence and female-guarding as male time investment strategies. *Behaviour* **48**, 156−184. (168)

PARKER, G. A. (1974b) Assessment strategy and the evolution of fighting behaviour. *J. theor. Biol.* **47**, 223−243. (76, 78)

*PARKER, G. A. (1978) Selfish genes, evolutionary games, and the adaptiveness of behaviour. *Nature, Lond.* **274**, 849−855. (1, 29, 73, 197)

PARKER, G. A. and STUART, R. A. (1976) Animal behavior as a strategy optimizer: evolution of resource assessment strategies and optimal emigration thresholds. *Amer. Natur.* **110**, 1055−1076. (168)

PARTRIDGE, L. (1978) Habitat selection. *In*: Krebs, C. J. and Davies, N. B. (eds) *Behavioural Ecology: an evolutionary approach*. Blackwell, London. pp. 351–376. (208)

*PAYNE, R. and WEBB, D. (1971) Orientation by means of long-range acoustic signaling in baleen whales. *Ann. N. Y. Acad. Sci.* **188**, 110–141. (67, 70)

PEARRE, S. (1973) A model of plankton migration. *Ecology* **54**, 300–314. (198)

PEARSON, K. and BLAKEMAN, J. (1906) Mathematical contributions to the theory of evolution. XV. A mathematical theory of random migration. *Drap. C. Res. Mem. Biom. Ser.* **3**, 54 pp. (5)

PERDECK, A. C. (1958) Two types of orientation in migrating starlings *Sturnus vulgaris* L. and chaffinches *Fringilla coelebs* as revealed by displacement experiments. *Ardea* **46**, 1–37. (8, 92, 149)

PERDECK, A. C. and CLASON, C. (1974) Spontaneous migration activity of chaffinches in the Kramer cage. *Progress Report 1973, Institute of Ecological Research, Royal Netherlands Academy of Arts and Sciences.* pp. 81–82. (209)

PHILLIPS, J. B. and ADLER, K. (1978) Directional and discriminatory responses of Salamanders to weak magnetic fields. *In*: Schmidt-Koenig, K. and Keeton, W. T. (eds) *Animal migration, navigation and homing* Springer, Heidelberg. pp. 325–533. (120, 128, 136)

PILLERI, G. (1979) The blind Indus Dolphin, *Platanista indi. Endeavour* **3**, 48–56. (117)

PRESTI, D. and PETTIGREW, J. D. (1980) Ferromagnetic coupling to muscle receptors as a basis for geomagnetic field sensitivity in animals. *Nature, Lond.* **285**, 99–101. (120)

QUINE, D. B. and KREITHEN, M. L. (1981) Frequency shift discrimination: can homing pigeons locate infrasounds by Doppler Shifts? *J. Comp. Physiol.* **141**, 153–155. (117)

QUINN, T. P. (1980) Evidence for celestial and magnetic compass orientation in lake migrating sockeye salmon fry. *J. Comp. Physiol.* **137**, 243–248. (120, 212)

RABØL, J. (1970) Displacement and phaseshift experiments with night-migrating passerines. *Ornis. Scand.* **1**, 27–43. (8)

*RABØL, J. (1978) One-direction orientation versus goal area navigation in migratory birds. *Oikos* **30**, 216–223. (12, 13, 92, 211)

RAINEY, R. C. (1951) Weather and the movements of locust swarms: a new hypothesis. *Nature, Lond.* **168**, 1057–1060. (184)

RANDOLPH, S. (1977) Changing spatial relationships in a population of *Apodemus sylvaticus* with the onset of breeding. *J. Anim. Ecol.* **46**, 653–676. (206)

RAVELING, D. G. (1976) Migration reversal: a regular phenomenon of Canada Geese. *Science* **193**, 153–154. (88)

REBACH, S. (1978) The role of celestial cues in short range migrations of the hermit crab, *Pagurus longicarpus. Anim. Behav.* **26**, 835–842. (164)

RICHARDSON, W. J. (1978) Timing and amount of bird migration in relation to weather: a review. *Oikos* **30**, 224–272. (88, 90)

RICHTER, C. P. (1933) The effect of early gonadectomy on the gross body activity of rats. *Endocrinology* **17**, 445–450. (206)

RILEY, J. R. and REYNOLDS, D. R. (1979) Radar-based studies of the migratory flight of grasshoppers in the middle Niger area of Mali. *Proc. R. Soc. Lond.* B **204**, 67–82. (185)

ROSIN, R. (1978) The honey bee 'language' controversy. *J. theor. Biol.* **72**, 589–660. (64)

*RUSE, M. (1979) *Sociobiology: sense or nonsense?* Reidel, London. (29)

SAILA, S. B. and SHAPPY, R. A. (1963) Random movement and orientation in salmon migration. *J. Cons. perm. int. Explor. Mer.* **28**, 153–166. (11, 106, 115)

SCHAEFER, G. W. (1976) Radar observations of insect flight. *In*: Rainey, R. C. (ed) *Insect Flight* Symp. R. ent. Soc. London: No. 7. Blackwell, London. pp. 157–197. (173, 185)

SCHMIDT, J. (1922) The breeding places of the eel. *Phil. Trans. R. Soc.* B **211**, 179–208. (153)

*SCHMIDT-KOENIG, K. (1979) *Avian orientation and navigation.* Academic Press, London (120, 122)

SCHMIDT-KOENIG, K. and KEETON, W. T. (1977) Sun compass utilization by pigeons wearing frosted contact lenses. *Auk* **94**, 143–145. (133)

*SCHMIDT-KOENIG, K. and KEETON, W. T. (eds) (1978) *Animal migration, navigation and homing.* Springer, Heidelberg. (121)

SCHMIDT-KOENIG, K. and PHILLIPS, J. B. (1978) Local anesthesia of the olfactory membrane and homing in pigeons. *In*: Schmidt-Koenig, K. and Keeton, W. T. (eds) *Animal migration, navigation and homing.* Springer, Heidelberg. pp. 350–365. (129)

SCHMIDT-KOENIG, K. and SCHLICHTE, H. J. (1972) Homing in pigeons with impaired vision. *Proc. Nat. Acad. Sci. U.S.A.* **69**, 2446–2447. (133)

SCHNEIRLA, T. C. (1971) *Army ants: a study in social organization.* Freeman, San Francisco. (63)

SCHÜZ, E. (1951) Uberlick uber die Orientierungsversuche der Vogelwarte Rossitten (jetzt: Vogelwarte Radolfzell). *Proc. XIII Intern. Ornithol. Congr. (Upsulla, Sweden)* pp. 249–268. (210)

SHARROCK, J. T. R. and SHARROCK, E. M. (1976) *Rare birds in Britain and Ireland.* Poyser, Berkhamstead. (91)

SHEVARYOVA, T. (1969) (Stability and changes of the nesting, moulting and hibernation places of waterfowl) (In Russian with English Summary). *Communs. Baltic Comm. Study Bird Migr.* **6**, 13–38. (222)

SIMMONS, J. A. (1977) Localization and identification of acoustic signals, with reference to echolocation. *In*: Bullock, T. H. (ed.) *Recognition of complex acoustic signals.* Dahlem Konferenzen, Berlin. (117)

SMITH, J. N. M. and SWEATMAN, H. P. A. (1974) Food searching behaviour of titmice in patchy environments. *Ecology* **55**, 1216–1232. (59)

SOTTHIBANDHU, S. and BAKER, R. R. (1979) Celestial orientation by the large yellow underwing moth, *Noctua pronuba* L. *Anim. Behav.* **27**, 786–800. (173, 203)

SOWLS, L. K. (1955) *Prairie ducks, a study of their behavior, ecology and management.* Stackpole, Harrisburg. (222)

STASKO, A. D. (1971) Review of field studies on fish orientation. *Ann. N. Y. Acad. Sci.* **188**, 12–29. (151)

*STONEHOUSE, B. (1978) *Animal marking: recognition marking of animals in research.* Macmillan, London. (24)

SYMONS, P. E. K. (1978) Leaping behavior of juvenile coho (*Oncorhynchus kisutch*) and Atlantic salmon (*Salmo salar*). *J. Fish. Res. Bd. Can.* **35**, 907–909. (97)

TAIMR, L. and KŘIŽ, J. (1978) Post-migratory local flights of *Phorodon humuli* Schrank winged migrants in a hop-garden. *Z. ang. Ent.* **85**, 236–240. (189)

TESCH, F-W. (1978) Horizontal and vertical swimming of eels during the spawning migration at the edge of the continental shelf. *In*: Schmidt-Koenig, K. and Keeton, W. T. (eds) *Animal migration, navigation and homing.* Springer, Heidelberg. pp. 378–391. (154)

THOMAS, A. A. G., LUDLOW, A. R. and KENNEDY, J. S. (1977) Sinking speeds of falling and flying *Aphis fabae* Scopoli. *Ecol. Ent.* **2**, 315–326. (189)

TRIVERS, R. L. (1971) The evolution of reciprocal altruism. *Quart. Rev. Biol.* **46**, 35–57. (73)

TRIVERS, R. L. (1974) Parent–offspring conflict. *Amer. Zool.* **14**, 249–264. (221)

TSVETKOV, V. I. (1969) (On the threshold sensibility of some freshwater fishes to the rapid change of pressure) (In Russian). *Vop. Ikhiol.* **9**, 715–721. (117)

TUCKER, D. W. (1959) A new solution to the Atlantic eel problem. *Nature, Lond.* **183**, 495–501. (153)

TUSKES, P. M. and BROWER, L. P. (1978) Overwintering ecology of the monarch butterfly, *Danaus plexippus* L., in California. *Ecol. Ent.* **3**, 141–153. (178)

URQUHART, F. A. (1976) Found at last: the Monarch's winter home. *Natl. Geogr.*, **150**, 161–173 (178)

UVAROV, B. P. (1921) A revision of the genus *Locusta*, L. (=*Pachytylus* Fieb), with a new theory as to the periodicity and migrations of locusts. *Bull. ent. Res.* **12**, 135–163. (181)

WALCOTT, C. (1978) Anomalies in the Earth's magnetic field increase the scatter of pigeon's vanishing bearings. *In*: Schmidt-Koenig, K. and Keeton, W. T. (eds) *Animal migration, navigation and homing.* Springer, Heidelberg. pp. 143–151. (117, 121, 136)

*WALCOTT, C. (1980) Magnetic orientation in homing pigeons. *IEEE Trans. on magnetics* **16**, 1008–1013. (120, 133, 135)

*WALCOTT, C., GOULD, J. L. and KIRSCHVINK, J. L. (1979) Pigeons have magnets. *Science* **205**, 1027–1029. (120)

WALLRAFF, H. G. (1966) Uber die Heimfindeleistungen von Brieftauben nach Haltung in verschiedenartig abgeschimten Volieren. *Z. Vergl. Physiol.* **52**, 215–259. (141)

WALLRAFF, H. G. (1974) *Das Navigationssystem der Vögel*. Oldenbourg, Munchen. (125)

*WALLRAFF, H. G. (1978) Preferred compass directions in initial orientation of homing pigeons. *In*: Schmidt-Koenig, K. and Keeton, W. T. (eds) *Animal migration, navigation and homing*. Springer, Heidelberg. pp. 171–183. (114)

WALLRAFF, H. G. (1979) Goal-oriented and compass-oriented movements of displaced homing pigeons after confinement in differentially shielded aviaries. *Behav. Ecol. Sociobiol.* **5**, 201–225. (141)

WALLRAFF, H. G. (1980) Does pigeon homing depend on stimuli perceived during displacement? I. Experiments in Germany. *J. Comp. Physiol.* **139**, 193–201. (128, 135)

WALLRAFF, H. G., FOÀ, A. and IOALÉ, P. (1980) Does pigeon homing depend on stimuli perceived during displacement? II. Experiments in Italy. *J. Comp. Physiol.* **139**, 203–208 (128)

WARD, P. and ZAHAVI, A. (1973) The importance of certain assemblages of birds as 'information-centres' for food-finding. *Ibis* **115**, 517–534. (70)

WHITE, H. C. (1936) The homing of salmon in Apple River, *J. Fish. Res. Bd. Can. N. S.* **2**, 391–400. (98)

WHITEN, A. (1978) Operant studies of pigeon orientation and navigation. *Anim. Behav.* **26**, 571–610. (140)

WIKLUND, C. (1977) Oviposition, feeding and spatial separation of breeding and foraging habitats in a population of *Leptidea sinapis* (Lepidoptera). *Oikos* **28**, 56–68. (162)

WILKINSON, D. H. (1952) The random element in bird 'navigation'. *J. exp. Biol.* **29**, 532–560. (106, 115)

WILLIAMS, C. B. (1930) *The migration of butterflies*. Oliver and Boyd, Edinburgh. (11, 184)

WILLIAMS, C. B. (1958) *Insect Migration*. Collins, London. (6, 11, 184)

WILLIAMS, T. C. and WILLIAMS, J. M. (1970) Radio tracking of homing and feeding flights of a neotropical bat, *Phyllostomus hastatus*. *Anim. Behav.* **18**, 302–309. (136)

WILLIAMSON, K. (1955) Migrational drift. *Proc. 11 Int. Ornithol. Congr. Basel, 1954*. pp. 179–186. (90, 144)

WILSON, E. O. (in May 1979) (222)

WILTSCHKO, R., WILTSCHKO, W. and KEETON, W. T. (1978) Effect of outward journey in an altered magnetic field on the orientation of young homing pigeons. *In*: Schmidt-Koenig, K. and Keeton, W. T. (eds) *Animal migration, navigation and homing.* Springer, Heidelberg. pp. 152–161. (129)

WILTSCHKO, W. and WILTSCHKO, R. (1976) Interrelation of magnetic compass and star orientation in night-migrating birds. *J. Comp. Physiol.* **109**, 91–9 (123)

*WILTSCHKO, W. and WILTSCHKO, R. (1978) A theoretical model for migratory orientation and homing in birds. *Oikos* **30**, 177–187. (127, 128, 142, 145)

WYNNE-EDWARDS, V. C. (1962) *Animal dispersion in relation to social behaviour.* Hafner, New York. (73)

Subject and organism index

Species are indexed under their scientific name. Groups (e.g. birds) are indexed under their common name.